SOCIAL WARMING

SOCIAL WARMING

THE DANGEROUS AND POLARISING EFFECTS OF SOCIAL MEDIA

CHARLES ARTHUR

A Oneworld Book

First published by Oneworld Publications in 2021

A CIP record for this title is available from the British Library

ISBN 978-1-78607-997-8 (hardback)
ISBN 978-0-86154-229-1 (trade paperback)
eISBN 978-1-78607-998-5

Typeset by Hewer Text UK Ltd, Edinburgh
Printed and bound in Great Britain by Clays Ltd, Elcograf S.p.A.

Oneworld Publications
10 Bloomsbury Street
London WC1B 3SR
England

For Saskia, Harry and Lockie

CONTENTS

1

PROLOGUE: THE SHAPE OF THE PROBLEM

Nobody meant for this to happen. Everything was meant to get better, not worse.

In January 2007, Steve Jobs, Apple's then chief executive, held aloft a little device in his hand. 'This is a revolution of the first order, to really bring the real internet to your phone,' he announced.[1] Until then, the internet had largely been confined to PCs; only a few million people had an internet-capable phone, and even they had limited capacity for viewing or interacting with online content.

Social networks were in their infancy. Facebook had twelve million users, having just opened up to the world beyond US university students the previous summer, about the same time as it had patented the software for a 'News Feed' that would pick out the most interesting status updates from your friends. Twitter was less than a year old and had tens of thousands of users.[2] YouTube had been bought two months earlier by Google for $1.65 billion – a price seen as astonishing, despite the site's estimated 70 million monthly users.

Mark Zuckerberg, founder of Facebook, originally defined its purpose as 'To give people the power to share and make the world more open and connected.' He tweaked that slightly a few years later, to 'give people the power to build community and bring the world closer together.'[3] The broad sweep was clear: to get people to communicate with each other more easily and directly.

In just over a decade, the world has gone from a time when barely anyone owned a smartphone to one where more than 4 billion of the world's 7.6 billion do, and almost all of them connect to social networks.[4] Walk today down a street past a construction site, past a coffee shop, past a parent pushing a pram, and you'll see the same thing: people gazing down at their smartphones, flicking through screenfuls of posts, photos, videos and comments. Smartphones have replaced cigarettes as the perfect filler for those empty moments, waiting for trains, meals or a friend. Pull it out of your pocket, watch it light up, suck it in and relax.

Just as smokers can measure their cigarette consumption in packs, we can measure our social media consumption in screens: in 2017, a Facebook executive said the average person scrolls through 300 feet of social media feeds per day on their phone. That's about 750 screens' worth, consumed over an average of two hours per day.

Social media as a crutch may be akin to smoking – something to do with our hands that also steadies our minds. But the cumulative effect is much more akin to global warming: pervasive, subtle, relentless and, most of all, caused by our own actions and inclinations.

That we use social media to fill our downtime is not a problem in itself; few would be reading *War and Peace* instead. But this use is closely monitored, the experience is individually tailored, and herein lies the harm.

Since social networks became widespread, Facebook has been implicated in a genocide, Twitter became the battleground for a misogynistic campaign leading to serious real-world threats and attacks, and YouTube has been accused of enabling the radicalisation first of Muslim jihadis and then of right-wing white men who would go on to kill. Women have received death and rape threats for campaigns about a banknote; football stars have been targeted for anonymous racist abuse from twelve-year-olds; and two men, who were brought together by an algorithm that spotted they were interested in the same topic, started communicating in a Facebook Group, and decided to act.[5,6] The topic was causing a civil war, and the action was to kill a police officer.

These aren't aberrations. Social networks have these results when used as intended, as designed. After all, you're *supposed* to connect with like-minded people.

The same pattern of events keeps happening when social networks are involved: small differences are amplified into bigger disagreements, and the people on either side of those positions are drawn towards extremes of belief or action. These networks are optimised to consume our attention, and powered by software that feeds on, and exploits, our inherent tendencies towards outrage and polarisation.

As long as social networks stick to their current design, events like these will keep happening, and get worse as the number of people using those networks increases. And in the next five years, another billion people will be able to access a smartphone.[7]

We're living in an age of 'social warming' – a side effect, an unintended consequence of technological advance making our lives more convenient.

We call it 'warming' because it's gradual. Gradualism means we don't quite notice the point at which things shift for the worse.

Social change isn't marked by abrupt shifts, but by almost imperceptible changes in behaviour and habits that are only obvious in retrospect. To take a trivial example, films and photos from the 1940s show almost all men wearing formal hats outside (which they raise to passing women), and everyone seems to be smoking. Nowadays, men don't wear formal hats, and hardly anyone smokes.

But there was never a single moment when men suddenly stopped wearing hats. Doing so just became less common as more people rode in cars, where a hat was an inconvenience, and as younger public figures such as John F. Kennedy, who never wore one, and the Beatles, who would never have dreamed of it, came to prominence. (Nor has the male need for a head covering in cold weather gone away. The formal hat industry mutated into the baseball cap and beanie industry.) Social warming arises from the desire to have a computer, the smartphone, that's allied with our hunger for information and desire to connect with more and more people. Its effects have only become noticeable as the adoption and power of social networks and smartphones has grown large enough to begin shifting our behaviour significantly.

Social warming happens when interactions between people who used to be geographically separated and infrequently exposed to each other's views are more frequently brought together, and kept in orbit around topics that will engage them and create addictive experiences.

Only when you look back does the change become obvious. The effects occur, though, all the time. The political sphere, democracy, media, people in the street: all are being affected.

Social warming comes about through a three-way interaction. First is the parallel rise of smartphone availability and social network accessibility. Second, each platform is able to learn and amplify what captures our attention, getting us to log in more frequently and for longer. Third, the amplification is unregulated and unrestricted. Partly this is by design – people using the system is good for business – and partly it's by management fiat, through a proscription against 'censorship'.

This repeated process of ubiquity, amplification and indifference, and its continuation, defines social warming. Without the platform, it couldn't happen. Without the amplification, we wouldn't notice it. Without the indifference to the effects, we wouldn't be exposed to them. And if it didn't keep happening, we wouldn't be so concerned.

Yet there are signs that the wider public is aware of what's happening; that we glimpse it out of the corner of an eye. We know it's there, yet can't quite catch sight of it.

In May 2020, the UK NGO Doteveryone, which aims to get the whole of the UK connected, published its final Digital Attitudes Report, looking at people's attitudes to technology. Among the findings was this odd fact: people thought the internet was better for them *as individuals* than for society as a whole. The gap was large: 80 percent felt the internet had made life a little or a lot better for them, but only 58 percent felt that the internet had had a positive effect on society. The gulf in attitudes was unchanged from the first version of the study, carried out two years previously. What did that tell us? 'People say, "I like the convenience of being able to do online shopping, but I worry that my high street is suffering as a result,"' Catherine Miller, then Doteveryone's interim chief executive, told me. 'I think there's a sense that you get immediate personal gratification from these services through technology, but you see the societal impact. There's not a direct line between me doing my shopping in my pyjamas at two in the morning and my high street looking sad and shabby. But I think there's a sense that the accumulated impacts on society are more obvious than the negative impact on individuals.'

Miller points to how conflicted we feel over this, even when it comes to social media: 'This is our infrastructure. I could try and delete Facebook, but then I wouldn't know where my children's football match is taking place this weekend. My partner boycotts WhatsApp, which is a source of intense irritation to me because it means that I get all the messages about where the football match is taking place, that I then have to copy and text to him so he knows which pitch to turn up at.' Her partner's boycott is a principled one, she says: 'He doesn't like the business, he doesn't like Facebook, he doesn't like Zuckerberg, he doesn't want to be part of it. But,' she adds, 'I think that's a fairly niche view these days . . . If your focus is on the social media aspect of things, I think it really is important to recognise the lack of meaningful choice.'

Ben Grosser, an artist and professor at the University of Illinois, points out that the companies rely on keeping us hooked – because otherwise they would cease to exist: 'These companies have no value without people donating their time and their media to the system,' he told me. 'So, ultimately what matters to Facebook, what matters to Twitter, at least what matters to their shareholders is that there's an endless stream of users, ideally an always increasing number of users, who are staying on the platform as much as possible, putting content into it. That content insertion then produces the data they can use for advertising.'

You want to escape. But you can't. Even if you don't directly contribute to social warming, everyone around you does.

If you had told Gottlieb Daimler or Rudolf Diesel in the 1890s that their designs for fuel-driven engines would in a little over a century's time be held responsible for rising sea levels, catastrophic hurricanes and the forced migration of millions of people, they'd have struggled to believe you. Their intent was honest and simple: they wanted to build efficient machines that would be used by people to improve their lives. The steam engines of the time were horrendously wasteful, burning coal and belching out smoke, with a fuel efficiency of less than 10 percent; petrol and diesel were more than twice as efficient. How could using less fuel be a bad thing? How could democratising transport and making it more widely available be wrong? 'The automobile engine will come, and then I will consider my life's work complete,' Diesel once said.[8]

The inventors of the internal combustion engine's modern equivalents – the social networks – have similar Pollyanna-ish aims. Facebook aimed to 'connect everyone'. YouTube promised to let you 'broadcast yourself'. Twitter would 'give everyone the power to create and share ideas'. But embedded in the systems behind each slogan was the mechanism to fascinate, outrage and eventually antagonise.

That third effect matters. Social warming shows up as polarisation, whether political or cultural. It's a sort of social 'heat', creating the potential for friction in any interaction with someone you don't know, whether in person or online (but particularly the latter), and with those you do too. Many people have had the experience of discovering that a relative is perfectly happy to be racist on Facebook, and to spew misinformation that you'd never expect them to utter face-to-face. Polarisation isn't good for society, because it creates barriers to the collective action that can benefit everyone. A classic example was the reaction in American states in 2020 to health measures that would reduce the potential for coronavirus infection. Because the public health discussion became polarised across party lines, some areas and groups ignored health advice about lockdowns and mask-wearing. People died who might otherwise have lived.

Yet it's hard to intuit a connection between retweeting a snide remark or angry headline and a country where half the population are unwilling to wear something as a public health measure, just as it's hard to make the connection between driving a car to the shops a mile away and the melting of Greenland's ice sheet.

Societies function best when they have common aims that bring people together: despite their destructive effects, natural disasters and wars provide a common goal for which differences are put aside. But social networks are built around division. They amplify differences by allowing every tiny variation in belief or interest to take on a life of its own. Even more, the dynamics of self-selecting online groups will drive them further and further away from common ground with other groups whose views differ even slightly from their own. Rather than providing a medium for societies to unite, social networks actually work in the opposite direction by giving everyone a way to discover their differences. That is social warming: the background effect that gradually, subtly, insistently makes people concentrate on their differences rather than what they have in common.

But wasn't it *always* like this? Isn't online interaction always more heated than real life, and nothing comes of it? Usually. Except when people threaten to kill MPs, or someone radicalised by a stream of videos picked for them by the software that powers a site goes on a shooting spree against their chosen enemies – a race, a religion, anything. At that point, something has evidently changed, and the online world, where 'things don't matter' and 'it's all just words on a screen', is bleeding into the offline one, where you really can drop things on your foot.

Our phones and social media identities have become our virtual homes. When a virtual mob begins targeting you, the effect isn't like being on a football pitch. It's not a wall of unintelligible slurs. Every insult on social media is isolated. It's as if each member of the mob were whispering in your ear. The suggestion that you 'just delete your account' or 'just ignore it' is the same as suggesting that you move home, or stay indoors.

We cannot ignore these effects, because they will not sort themselves out. Facebook and Google can be used to swing elections. Facebook is proud of its ability to persuade millions of voters to register and even to turn out to vote; yet one of its executives, Katie Harbath, was also prepared to accept that the election of Rodrigo Duterte in the Philippines in early 2016, following a brutal social media blitz of misinformation and personal attacks on opponents, made that country 'patient zero' in electoral interference through social media. (She then went on to cite the Brexit referendum in the UK and the 2016 US presidential election as other examples.)

The side effects of social networks grow geometrically faster than the networks themselves. But the legislative systems they've effectively encircled can't respond at the same speed. Legislators work over periods of years, while social networks can roll out new updates in weeks or months. By the time a committee of members of parliament in the UK came to consider the problem of 'fake news' in January 2017, the 2016 US presidential election and Brexit referendum that had made the topic urgent had long since passed, and a tweak to Facebook's and Google's software had pushed the problem out of sight for most people. The committee was then dissolved by an election; the final report appeared in July 2018. No laws were passed.

Social network companies are reluctant however to take ownership of the consequences of their choices. They're happy to take credit for the positive effects, such as when people can 'check in' on Facebook to confirm they're alive after a natural disaster, or activists can use Twitter to record wrongful arrests, or you can find the instructional video for fixing your lawnmower on YouTube.

Yet when they help Nazis and provocateurs to organise into closed groups, enable harassment, or send vulnerable people down rabbit holes of conspiracy theories, their response is apologetic and puzzled: 'How did *that* happen?' they ask. The downsides – what economists call the 'negative externalities' – become a problem for society to deal with and pay for, even though the software-driven amplification of outrage and interaction caused those effects in the first place.

Nor is there any clear way to bring external pressure to bear on the networks to make them directly answerable for those effects. Facebook and Google have corporate structures in which their chief executives and founders hold a majority of the voting shares, insulating them from shareholder ire. Literally the only person who can remove Mark Zuckerberg from his position at the top of his company is Mark Zuckerberg. The only shareholders Larry Page and Sergey Brin answer to at Google, and hence YouTube, are themselves: they own about 80 percent of the voting stock. (Twitter has a more straightforward ownership structure, where public shares have voting rights equal to founders'.)

Looking ahead to where those new mobile internet connections and smartphones will be found in the next five years, almost all will be in less developed countries in regions such as sub-Saharan Africa and Latin America, where weaker democratic and media systems will find it harder to withstand the onslaught of untruth and distortion. What then happens to democracy? What happens to truth? What happens when a population can't even agree about what happened a day or a month ago, or who won an election, and those disagreements are reinforced every time they look at the device in their hands? Or what about when it's cheaper and easier to get misinformation than to get facts, as is the case in a number of countries where mobile carriers offer deals that make access to Facebook or WhatsApp free, but access to a search engine or news site paid-for? This is an emergency: it needs to be tackled by recognising the toxic effects and removing the elements that enable them – the incitement to outrage, the algorithmic nudging, the pretence that throwing everyone into one giant room and encouraging them to shout at one another, or even flatter one another, will make them happier in the long term.

Antonio García Martínez, who worked on Facebook's most successful efforts to make money from advertising, noted in his book *Chaos Monkeys* that even a little bit of difference is worthwhile if your network's big enough. To the criticism that any individual Facebook advert didn't bring in much money, he responded that 'A billion times any number is still a big fucking number.' We should be a lot more worried than we currently are about these big fucking numbers. Only then do we stand a chance of figuring out what to do.

2

EARLY DAYS: THE PROMISE AND THE POWER

I will build a car for the great multitude ... no man making a good salary will be unable to own one, and enjoy with his family the blessings of hours of pleasure in God's wide open spaces. – Henry Ford, 1903[1]

From the moment computers became available to the general public, people began creating social spaces online. Usually called a 'bulletin board' or BBS (bulletin board system), because the format mimicked the communal boards in an office – you post your notice, people come and read and perhaps write on it, and others respond – BBSes quickly demonstrated the particular ways in which online interaction could differ significantly from the physical form. Notably, that you could be a lot ruder or more untruthful than you might be in real life without suffering any particular sanction.

One of the oldest bulletin boards is The WELL, which started in 1985 in Sausalito, California. The name is an acronym for Whole Earth 'Lectronic Link, retrofitted because one of the creators had created the *Whole Earth Catalog*, a printed magazine. It attracted the vanguard of internet utopians; rather than being set up as a money-making enterprise, The WELL was intended as an experiment in what would happen if you let people communicate unmediated in a big group online.[2] One of the co-founders,

Stewart Brand, also wanted to encourage users to meet face-to-face, but that wasn't compulsory. A key choice was an insistence on using real names, banning anonymity. 'You own your own words', the site's motto read. Brand later said he had been trying to foresee, and so forestall, what might go wrong in such a space: 'One thing would be people blaming us for what people said on The WELL,' he recalled in an interview with *Wired* in 1997. 'And the way I figured you get around that was to put the responsibility on the individual.'

Access to The WELL wasn't free, but Brand and co-founder Larry Brilliant tried to set it as low as they could for the time: that turned out to be $8 per month for membership and $2 per hour for access. Such prices seem extortionate today; then, they were bargain-basement.

A few things about The WELL's discussion system would become axiomatic for almost all future systems. Postings in the discussions (called 'conferences') didn't expire; anyone could reply to publicly visible posts, though some conferences could be made private so only invited people could see them; and deleting posts was difficult. (Deleted posts left a placeholder indicating who had created and deleted it.) The posting software had a steep learning curve that automatically divided users along lines of expertise and, once they'd mastered that, typing speed – for even in the later years of the twentieth century, typing was not a common skill. Partly for that reason, and partly because of the location, quite a number of the early users were journalists or computer technicians, whose jobs already involved banging keys and who were likely to have a computer.

Among the journalists who became enthralled with the community on The WELL was Howard Rheingold, who found himself sucked in when his first post (about tarantula sex) was eagerly received: 'you know your behaviour is somehow obsessive and taboo in the Protestant sense, that you should be working . . . but you also know that it's sociable, and you're doing it together,' he told *Wired*.

But to create paradise is always to ask for trouble. That came in 1986 with a new WELL user who chose the screen name Mark Ethan Smith, but was actually female, and would insult and roar virtually at people who disagreed with claims – many of them demonstrably wrong – that she made about feminist history.

She didn't, however, get thrown off the site. Instead, Matthew McClure, whom Brand had hired as The WELL's director, decided that 'Smith' was playing with the users' cultural expectations; that she understood how they would react better than they did, and 'just played it like an instrument'.

Smith also generated a lot of attention and argument from other users, which meant login time, which meant revenue. That mattered to The WELL, which was losing money. Even so, Smith eventually proved too much of a troublemaker; the extra revenue she brought in didn't counterbalance the ire she generated, and her account was terminated in late 1986. Smith has described this as 'censorship in cyberspace' in a long personal history posted online, claiming the title of being the first person to be kicked off The WELL, describing the ban as 'a vicious and unconscionable act of censorship' and an 'abrogation of my freedom of speech'.[3] Smith argues that what others had seen as verbal aggression was instead a personal response to their perceived aggression – particularly their use of female pronouns. Smith had effectively renounced gender after a number of work-related problems, and so responded in kind by referring to men as 'she', which often irked them.

Smith and The Well provided an early example of the inherent conflict that came to shape many social networks in the following years; having people who rile others is terrific for enhancing engagement, particularly if you make your money from how much time people spend on the site. (For The WELL, from charging for access; for later networks, from advertising.) Having users who outrage the rest sufficiently to make them keep coming back, yet not enough to make them swear off using it, is a surprisingly effective business model. Even the complaint of the evicted user is familiar: they are being censored; their freedom of speech is being interfered with. The implicit belief is that if someone else creates a platform to let people speak, then that automatically gives every user the right to use it in any way that they, not the owner, want.

One of the most important moments for the development of social networks was not a technical advance, but a legal case in 1995. Four years earlier a different provider, CompuServe, which operated a huge number of forums, had been sued for potentially libellous content on one of its forums. (A daily newsletter about journalism published there called a rival a 'new start-up scam'.) By insisting it did not moderate the forums' content, CompuServe successfully argued that it was a 'distributor' like a bookshop or telephone company, not a newspaper publisher, and so was protected under the law. The 1991 decision set a precedent for the internet.

Shortly thereafter Prodigy, an American ISP (internet service provider), was sued by an investment banking firm over anonymous claims of fraud made on one of its forums. Prodigy offered the same defence as CompuServe. But it lost because, crucially, both humans and software moderated its forum content; that meant it was not like a shop, but more like a newspaper.[4] The liability from losing the case ran to millions of dollars. The implication was clear: don't moderate forums, or else you'll be liable. Yet being unable to remove content for fear of liability would mean forums could turn into a mass of illegal content – spam, libel, stolen software – which would put off ordinary users, and create huge downsides that could undermine the burgeoning internet business.

ISPs lobbied US senators who were then considering the 1996 Communications Decency Act, a huge new bill being pushed by the new Clinton administration. It had been prompted by one of the periodic spasms of puritanism in the American national psyche about the possibility of pornography finding a new outlet (in this case, the internet). The CDA's initial draft made it an offence to 'knowingly' send indecent or obscene material to minors. If that became law then ISPs would have to filter content – but the Prodigy decision would also make them liable for any libels or other infringements by their customers. Nobody would run an internet business in the lawyer-heavy US given that double bind.

Discussion on the internet, at least in the US, was saved by a bipartisan duo of senators, the Democrats' Ron Wyden and the Republicans' Chris Cox. They drafted a clause – Section 230 – to add to the CDA. It achieved the seemingly impossible, absolving companies of immediate liability for what was posted on their forums while simultaneously allowing them to moderate content as they liked. 'It's this two-sentence thing, which is basically a Get Out Of Jail card,' explains John Naughton, a Cambridge University professor and author of *A Brief History of the Future: The Origins of the Internet*. 'It says that if you're just hosting things, you are not responsible for what people do on it. That's the key moment: that's why these huge companies have grown, on the basis that they're not responsible for what happens on their platforms. They're not legally liable for it. That's the key bit.'

A year later, the US Supreme Court overturned the part of the CDA relating to indecent content on the basis that it violated freedom of speech, effectively gutting the 'Decency' part of the act. But Section 230 survived, and would underpin the ability of providers to let people post what they wanted without having to check it first for legality.

Without Section 230, there would be no Facebook, no Twitter, no YouTube. There would probably be a lot of lawsuits, and the web would largely consist of scientific papers, which it was initially designed to connect, and lots of bland corporate sites. (And, surely, pornography, at least outside the US.)

Instead, Section 230 meant that while the writer might have to own their words, the site that hosted them didn't have to. Sites could remove content as they liked, but weren't liable even so for what they left. The 'Good Samaritan' clause, §230(c), conferred legal immunity for 'any action taken in good faith to restrict access to or availability of material that the provider or user considers to be obscene, lewd, lascivious, filthy, excessively violent, harassing, or otherwise objectionable, whether or not such material is constitutionally protected'.[5] (The thesaurus-level focus around lewd tells you a lot about the clause's origins in the CDA.)

Two things are important about that clause. First, providers don't *have* to moderate; if they want to host 'lewd, lascivious, filthy' or 'excessively violent' content, they can (though obscene and illegal material, including child abuse material, would never be allowed). Second, the final clause about 'constitutionally protected' material short-circuits any complaint that platforms that moderate are infringing the US Constitution's First Amendment, which bans the *government* from preventing speech and gives citizens wide-ranging rights to speak. Instead, it asserts that internet platforms are the property of the companies that run them, to do with as they please. Section 230(c) meant that the complaints of censorship by Mark Ethan Smith's successors would be just as hollow in the future as the original one had been. 'It's a piece of legislation which has determined everything that's happened since,' says Naughton. 'You can see why Wyden and Cox thought this: they figured that if they don't have this two-sentence clause, then this thing is going to be screwed. It's not going to grow because every goddam lawyer in the country will be onto it, and nobody will be able to do anything. You can see it was a wise decision at the time. But it has gone to places that nobody could have forecast. And now we're living with the consequences of it.'

During the internet's early years, 'social' discussion online took place first on mailing lists, and then on web pages in forums where you could write 'posts' with your contributions. Forums used simple software that organised the discussion on any topic with the oldest entry first, and everything afterwards following a chronological layout, with reactions and responses strictly ordered by the time of posting. The index page to all of the forum posts, however, was organised in reverse chronological order, so that the topics and discussions with the most recent updates appeared at the top.

Blogs ripped up that idea. A blog – short for 'weblog', a log (journal) on the web – was intended to be a regular update by one or more people about whatever they liked. But nobody would want to slog through a forum-style mass of old content to find the new stuff. So, blogs were organised to show the most recent posts first, and only the determined would scroll on to older material. The past became a different, optional country; all that mattered was the present. Blogs were also different from most of the web before: they were often intensely personal, written in the voice of someone speaking conversationally, not formally.

The very first blog appeared in 1994, and in 1999 a company called Pyra Labs released a template that would become known as Blogger. People without arcane technical skills could quickly set up a blog, and then compose blogposts by logging into a website, writing something, and hitting a button marked 'Publish'. The truly democratic age of the web had begun: anyone could have a voice.

In 2003, Google bought Pyra Labs and offered Blogger blogs for free, monetising them with ads injected onto the page based on the content of each post. The same year, Wordpress, a comprehensive package of free code for running a blog, was released.

The combination of free software and free blogs triggered a Cambrian explosion of self-expression. The number of blogs grew rapidly, from just over twenty in 1999 to more than fifty million by the middle of 2006, according to Technorati, a company set up to follow activity in this exciting new space.[6] Exponential growth ruled: for three years, the number of blogs doubled every six months. The sky seemed to be the limit.

Unlike forums, blogs weren't democratic. They gave the author complete dominion to push their own voice and opinions. Blogs revolved around the ego, unlike forums, where the voice of the crowd ruled supreme. (Reddit is essentially a topic-based forum: you join in order to participate in a discussion around a subject, rather than to converse with specific people. In that sense, Reddit isn't a social network, and so isn't treated as such here – even though many of the deliberations that happen there spill over into Twitter and Facebook. Reddit is gigantic, but with a primary focus on topics, not the individual.)

With the arrival of blogs, the internet appeared once again to have created a paradise for discussion. And once again, a snake emerged. Unlike The WELL, the sin this time was not wrath, but envy. Even though lots of people had started blogging at roughly the same time, a comparatively small number of bloggers seemed to get most of the attention and readership. What sort of unfair system was at work? What was being rigged so that only a few were famous?

By the time a long article in the *New Yorker* in November 2000 brought the word 'blog' to the attention of metropolitan readers, the differential in visibility had already grown large enough that snarky posts complaining about the 'A-list' were common.[7] 'It's not that you missed the Golden Age. It's just that the age is golden only for other people,' wrote Joe Clark, a Toronto-based journalist musing on the *New Yorker* article's identification of both a hierarchy and an upper clique of bloggers.[8] 'And there is pretty much no way to breach the velvet rope: if you're not an A-list blogger, you will stay off that list forever . . . It bugs me that the A-list kids are not really any smarter, or any better at Web design, or have anything particularly better to say than so many of the plebes. Their fame is inexplicable, but famous they are – and able to keep their heads above water.'

As the number of blogs exploded, that A-list effect remained in place: the number of readers going to the most-read blogs was far bigger than for those on the next tier, which in turn was far more popular than the next, and so on until you hit the long tail of millions of blogs where often people had tried a few posts, received no engagement and given up, leaving unattended pages that formed a sort of cosmic background hiss of blogging.

What were the A-list bloggers doing that was so different?

There was indeed a new force at work. Once networks can effectively spread without limit, and effectively without cost – and the growth of blogs from around zero to fifty million in six years was a classic case – then a different force takes over. We're used to hearing in everyday life about 'bell curves', also known as the 'normal distribution': the height of the people in a population, babies' weights at birth, reaction times. Forget about them. 'There are no normal distributions in this technology anywhere,' says Naughton. 'What you find is that almost everything is governed by the power law.'

The power law, also known as Zipf's Law, is brutal. Though not much taught in school, it's surprisingly common in life. It's often known as the 80–20 rule: 80 percent of people have 20 percent of the wealth, and vice versa. In a set created by the power law, the value of the item in the Nth position is 1/N. If it were income, then the person in the first position might get $1 million, the second $500,000, the third $333,333 and so on. By the time you reach the hundredth person, they're getting $10,000. There are plenty of real-world examples of power laws: the size distribution of villages and cities (plenty of the former, few of the latter), the cost of individual insurance claims, the frequency of word use, the size of sand grains, incomes in the US, and popularity on the dating app Tinder, where 80 percent of women compete for the most desired 20 percent of men.[9]

In 2003, Clay Shirky, a professor at New York University who was deeply involved in the early expansion of blogging, wrote a post on his personal blog called 'Power Laws, Weblogs, and Inequality'. He pointed out that the complaints among bloggers about an 'A-list' had been heard before whenever a new social system started up, including The WELL. 'A new social system starts, and seems delightfully free of the elitism and cliquishness of the existing systems,' he noted. 'Then, as the new system grows . . . Some core group seems more connected than the rest of us, and so on.'

The explanations, or complaints, that those at the top of the popularity game had somehow cheated or sold out were wrong, Shirky explained. 'What matters is this: diversity plus freedom of choice creates inequality, and the greater the diversity, the more extreme the inequality.'

Shirky showed that the pattern of links to 433 of the larger blogs at that time followed a power law, as did the number of subscribers to a number of mailing lists, and links between blogs. 'Power law distributions tend to arise in social systems where many people express their preferences among many options . . . [and] as the number of options rise[s], the curve becomes more extreme.' This might seem counterintuitive – wouldn't having more choices mean people would range far and wide? – but the evidence was that it didn't: 'increasing the size of the system increases the gap between the No. 1 spot and the median [middle item] spot.' That is, inequality (in the absolute sense) gets worse.

This had two knock-on effects. First, in a power law distribution, most items are below average, because the items at the top distort the numbers so wildly. (In a bell curve distribution, half of the items are above average, half below it.) In Shirky's example of links between the 433 blogs, two-thirds had fewer than the average number of incoming links. The number of links for the mid-placed, or median, blog – the 217th – was half the overall average. (For a normal distribution, the median is also the average.)

Second, a power law distribution is very difficult to disrupt, because there is a figurative mountain to climb in order to reach an influential position. New bloggers would have to try to gain attention somehow – perhaps by linking to one of the bigger blogs to get attention? Which would reinforce that bigger blog's position, entrenching the thing they were trying to dislodge. The A-list would be more secure.

As Shirky pointed out, 'changing this distribution would mean forcing hundreds of thousands of bloggers to link to certain blogs and to de-link others, which would require both global oversight and the application of force. Reversing the "star" system would mean destroying the village in order to save it.' He didn't have any good news for those looking to blog: 'It's not impossible to launch a good new blog and become widely read, but it's harder than it was last year, and it will be harder still next year.'

Within a couple of years, the explosion in personal blogging that had begun with Blogger and Wordpress fizzled, even though the format was increasingly adopted by all sorts of corporate organisations and governments as a method of getting their message out to the public. In theory, the internet was a level playing field – you could link to me, I could link to you, we could link to a third and a fourth and a fiftieth or 200th or 2,000th blog – but in reality it was tilted. A few people got lots of connections; lots of people got a few.

That didn't turn people off using the internet. It just left an unfulfilled gap: something that would make it easier for people to express themselves, while not becoming a popularity contest that they could only lose.

What filled the gap was 'social software' – what we now call social networks.

The phrase 'social software' seems to have first been used in 1987, in a paper written by Eric Drexler of the Foresight Institute that looked at what a hypertext publishing system might enable. (What we know now as the web was then still four years away.) 'Social software could facilitate group commitment and action ... The possibilities for hypertext-based social software seem broad.'[10]

Sites that were set up explicitly to connect people first began appearing in the late 1990s, principally for dating. But more general ones began to emerge. Though many people were still on slow dial-up connections, a growing number on the American West Coast had faster, always-on broadband connections. That changed their relationship with the internet, which became as quotidian as electricity or running water, rather than a specific place that required specific visiting arrangements – a screeching modem, a ticking clock on the cost of connection. The only smartphones, however, were incredibly clunky devices that struggled to sip data from painfully slow mobile networks that were even slower than the dial-up internet.

Social software described applications that allowed groups of people to 'communicate and collaborate'. The group element was what mattered; previous systems had mostly been one-to-one, or one-to-many.

There was plenty of mathematical theory about the behaviour of social networks, as Shirky had highlighted. But there was also social theory about why we might like them, and how we might behave given the chance to join them. René Girard, a French social studies theorist, had previously built a theoretical framework suggesting that humans are mimetic animals: through a process that begins in infancy with us observing adults, we seek in life the things that are wanted by other people whom we admire. In its simplest version, you admire a celebrity, and they publicly desire (and use) a particular brand of makeup, and so you want that makeup. But that leads to competition: others use it too, and for the same reasons. So you have to show that you're more worthy of using it than they are, and perhaps even more worthy of using it than the celebrity. The competition that ensues can be destructive, or pointless. Girard's ideas were used as an explanation of financial bubbles in which people want otherwise useless things, because they see others doing so. (Think of tulips, or bitcoin.) This is 'mimetic' – copying – behaviour.

If we saw other people joining a social network, Girard's theory would predict, then we too would want to join; and because we could use the network to observe others and find things of theirs to desire, we would find the experience even more gratifying.

One of the first social networks to gain widespread use in the US was Friendster, which launched in 2003 and soon had three million users, almost all in the US. Friendster and the social networks that followed let you blog without the mess of blogging, and particularly without the effort of finding and keeping an audience, since you had a ready-made circle of friends or followers. 'You can use Friendster to meet new people to date, through your friends and their friends,' its front page suggested, beside a graphic showing a number of people's faces connected in a social graph. You could 'Make new friends' or 'Help your friends meet new people.'

Observing the rapid growth of Friendster in October 2003, David Kirkpatrick wrote at Fortune.com that 'There may be a new kind of Internet emerging – one more about connecting people to people than people to websites ... In the explosive growth of social networking we are surely seeing the future, using the Net to connect people with bonds of trust and friendship – and maybe sex.'[11]

Friendster was quickly superseded by another rival, MySpace, which had also started in 2003. Others quickly followed: Ringo, Bebo, Path, Orkut, Foursquare, Pownce, Jaiku, Qaiku, Tribe, and many, many more as venture capital poured into the hot new sector of social networking, seeking the next Google – but for people rather than websites. In February 2004, a nineteen-year-old student in his second year at Harvard called Mark Zuckerberg launched a site called TheFacebook, intended to let the university's students and alumni keep in touch through a virtual directory that let them list personal details and message each other. Its rampant popularity on the campus led him to the abrupt realisation that the idea could be expanded further. 'I thought, "You know, someone needs to build a service like this for the world,"' he later recalled.[12]

One of Facebook's earliest investors was a fan of Girard's work, and thought the project was a validation of those ideas: 'Facebook first spread by word of mouth, and it's about word of mouth, so it's doubly mimetic,' he said.[13] 'Social media proved to be more important than it looked, because it's about our natures.' The investor was Peter Thiel, and his initial $100,000 investment became more than a billion dollars, a ten-thousandfold growth, when the company floated on the stock market.

Soon renamed Facebook and targeted at American college attendees, the site's growing popularity and the evident wider demand for 'social network' sites saw access opened up to everyone in 2006. The same year, a trio in Silicon Valley – Evan Williams, the original founder of Blogger, together with Biz Stone and Jack Dorsey – gave up on their idea of a podcasting start-up and pivoted instead to creating a social network that would use something like the status messages from AOL Instant Messenger, but that could function over the phone network. Messages would be limited to the SMS length of 140 characters, but that was enough for twenty words or so. 'just setting up my twttr,' wrote Dorsey in the first public message on 21 March 2006.[14] In October, Google swooped in to snap up a new video site called YouTube, which had been set up in 2005 as a dating site where would-be catches could upload videos of themselves. Like the Twitter trio, YouTube's founders had quickly pivoted away from their original idea, and let people upload anything. Google paid $1.65 billion, and killed off its own faltering video site.

In May 2009, Facebook's traffic passed that of MySpace, according to the measurement company ComScore.[15] By the end of that year, Facebook claimed more than 300 million users; MySpace had 100 million.[16]

Once Facebook passed MySpace in size and became the uncontested leader, the graveyard began to fill up with would-be rivals. Facebook's growing user base became a potent illustration of a phenomenon called the network effect: just as a telephone becomes more and more useful as more people overall have a telephone, Facebook benefited from the fact that as more people used it, more people who *weren't* using it wanted to join and find people to get in touch with, and the more likely they were to find them.

Only a few rivals were left standing, notably Twitter. In October 2008, Zuckerberg offered to buy it for $500 million – $400 million in Facebook's still-private stock and $100 million in cash. Twitter at the time was still known as a 'microblogging' service, because nobody could think of any way to describe short personal updates on the web besides 'blogging'. It had six million registered users, a number that had grown sevenfold since the previous year. Evan

Williams, the new chief executive who had replaced Dorsey earlier that month, rejected Zuckerberg's offer on the basis that Facebook's valuation, based on investment rounds, was inflated. Twitter would go public in November 2013, valued at $24 billion.

Even as the number of choices for social networks narrowed, public excitement grew about what this new hyper-connected world of social networking could mean. There was no shortage of forecasts. 'They could lead to ways of finding and interacting with one another we never imagined,' wrote Lisa Hoover, a journalist at *PCWorld*, in April 2009.[17] 'They are taking us somewhere exciting ... [they] expand the pool of people we have the opportunity to meet to near limitless possibilities.'

Sarah Gavin, director of global communications for Bebo (which was briefly the most-used social network in the UK, before selling itself to AOL in September 2008), told the BBC in September 2006: 'It's really powerful. I think it's the first time that individuals have got the power ... It's a hugely powerful medium and people are just starting to grasp how effective that can be.'[18]

Martin Stiksel, a co-founder of the music-based social site Last.fm, agreed: 'If there is a possibility to pool all this knowledge, like there is in a social network, to the benefit of everybody, that's a really, really powerful thing.'

In November 2009, Biz Stone, who had succeeded Williams through the revolving door of Twitter CEO, went even further. Social networks would make the world a better place simply by enabling more information to flow, he told a panel at the Reuters Institute in Oxford: 'On a large scale, the open exchange of information can even lead to positive global impact. If people are more informed they are more engaged, and if they are more engaged they are more empathic. They are global citizens, not just a citizen of a nation.'[19]

Zuckerberg seemed to tacitly agree with that notion, writing in a 2009 post on the Facebook blog that 'Our main goal at Facebook is to help make the world more open and transparent.'[20]

The only note of dissent came from Reid Hoffman, the founder and CEO of LinkedIn, who wondered at the same panel where Stone spoke whether we'd really seek out information to make the world better, or just more of the junk that we liked to consume in supermarket newspapers or celebrity magazines. 'You might think, "Who wants to consume all this useless information?" But with some information, it is like with ice cream,' Hoffman said. 'It is not nutritious, but people still eat it.'

Hoffman's warning, though, was a cloud no bigger than a man's fist amid the blue skies that social networks promised. Everyone could be everyone else's friend! Everyone could write tweets, and respond to others' tweets! The open, transparent world beckoned. But nobody, including Thiel, paid much attention to another element of Girard's theory about mimetic behaviour: that our constantly frustrated desires first to have the things we saw other people have, and then in effect to *be* those people, would repeatedly drive us to form angry mobs that would destroy the enemy – once the mob had agreed on precisely who or what the enemy was. He called it 'scapegoating'.

There was certainly evidence that people would behave in that way as the networks grew. Facebook was for its first few years thought of solely as a place to chat to your friends; Twitter as a place to, as its denigrators put it, 'tell the world what you just had for breakfast'. Social networks seemed to fit the description given to Planet Earth in Douglas Adams's book *The Hitchhiker's Guide To The Galaxy*: mostly harmless.

Signs of trouble emerged gradually. 'Trolls' were already a familiar breed to people who had spent time on early internet forums and on Usenet, the non-commercial topic-based system that in the 1990s foreshadowed Facebook's Groups by letting people read or post commentary on almost anything. Unlike Smith on The WELL, who was sincere but infuriating, trolls set out to annoy. They target those they see as overly sensitive or foolish and find ways to annoy or (ideally) enrage them, either by saying outrageous things or pretending to be stupid, and take delight in the result. Savvy users learned to recognise them; 'don't feed the trolls,' they would counsel those they saw being sucked in.

The widespread adoption of social networks offered unending supplies of fresh meat to trolls, and they fell on it eagerly. The surviving parents or relatives of people who had killed themselves were favourite targets: from the mid-2000s, newspapers began to fill with stories about 'cruel internet trolls' who had in some way defaced MySpace or Facebook 'memorials', particularly to dead children. Sometimes they were singular, often they were organised in groups who sought to expose and embarrass people. Creating fake documents or blogs for effect, hacking personal web pages, publishing targets' phone numbers and encouraging others to call them; the determined internet troll had a large toolbox to work from. The journalist Mattathias Schwartz suggested in a 2008 article that the urge to troll comes from something misanthropic – a hate of others – inherent to all humans, but only rarely acted on.[21]

Milder than trolling, but closely related, was the lack of inhibition that many discovered on going online. Presented with a screen and a blank space, many felt no need to hold back their opinions about what or whom they were responding to. Predictably, the quality of discourse would quickly deteriorate. Being anonymous led people to behave more antisocially; the effect was like swearing at other drivers from the safety of your car, protected from the outside world, but online, it was even more tempting. The process had been clearly documented in a much-cited July 2004 paper by John Suler, professor of psychology at Rider University in New Jersey.[22] He identified six factors at work: people felt disconnected from what they were writing;

the other participants were literally or figuratively invisible; any discussion was asynchronous, with no way to force responses; you had to decide for yourself how to interpret others' comments (as serious, ironic, foolish and so on); what you wrote wouldn't impact your physical life; and, finally, real-life status and authority didn't transfer onto a screen, so everyone was effectively equal, squashed into the same number of pixels on the screen. (Suler also dismissed the suggestion that how users behaved online revealed their 'true' identity. For someone who was shy in person yet outgoing online, neither was the 'true' representation, he said; they were just facets of the same persona.)

Sometimes people even seemed surprised by the difference between their own online and offline behaviour. 'People get sucked in,' one poster told MSNBC, in an article exploring the effects of online anonymity. 'You can be whoever you want, you can put out there whatever you want, and there are no consequences. I even got sucked in and was mean to people.'[23]

Only one thing was needed to push people's behaviour towards a less inhibited online persona: greater access to social networks, as the disinhibition effect of being able to say almost anything to anyone, and to garner an audience of unpredictable size, became at first refreshing and then intoxicating.

In the West, social networks first flourished on the desktop PC. This was, in retrospect, a strange environment for software ostensibly intended to be personal and light-touch. The big bright screen and multiple distracting windows weren't ideal for producing a product that would draw people in.

That began to change in 2007, with the arrival of Apple's iPhone and, a year later, the first Android smartphone from Google. They showed that you could take your leisure everywhere with you too. The plain mobile phone had already suffused the population, but now smartphones brought email (personal at first, and then work too), web browsing, video and soon a galaxy of apps.

Within a couple of years, millions more people began to understand the fascination of always being connected. As a technology journalist at the time, I was fascinated by the changes over the years in the technological landscape of my daily commute, which involved an hour-long train journey into London and then a few stops on the London Underground. In the mid-2000s, there would be a few earnest people tapping away at their laptops on the train, and many bored faces on the London Underground. As smartphones became cheaper and more functional, there was a gradual flowering of little screens in more and more hands. People looked at their smartphones a lot more than they had their iPods and other music players: the screen had a lot more going on, including games and connections to the social networks that were becoming increasingly popular. Smartphones were clearly becoming the biggest rival for attention time with newspapers, which in London meant both morning and late-afternoon publications. Only when underground, with no signal, did people consider picking up a paper, though it wasn't long before they were playing games on their phones' little screens, and papers became an afterthought.

The split in connectedness was no longer between work, home and commuting; instead there was 'with smartphone' and 'without smartphone'. Before long, the only situation where people expected to be 'without' was if they were asleep, out of battery or mugged.

That created the potential for new businesses, new ways of working, new ideas about what was possible. The last two big consumer software products aimed at desktop computer users were the file-hosting service Dropbox, launched in September 2008, and the music service Spotify, released as an invitation-only product in October 2008. After that, the breakthrough products were mobile apps. Instagram launched in October 2010 (and was inaccessible from the desktop until November 2012) and the cab-hailing service Uber a few months later. Services that didn't adapt to mobile usage floundered.

The arrival of widespread use of smartphones after 2010 also changed the calculus of social networks. When they were limited to PCs, there would be long hours during which people were away from a screen – commuting, eating, walking outside. The smartphone offered an entirely new territory of time and attention. Social networks set about the task of conquering it with alacrity, designing mobile apps that would notify you of the most minor detail – someone commented on an update of yours; someone else replied to a tweet – in order to grab your attention and hence time.

Plenty of people were certain that getting a smartphone into everyone's hands would be an unalloyed benefit. 'In our lifetimes we're going from almost no one being able to communicate to almost everyone be able to communicate,' said Eric Schmidt, then Google's CEO, in a wide-ranging interview in 2009 with Charlie Rose. 'We're also going from almost no one having any kind of information and access to libraries to virtually everyone having access to every piece of information in the world. That is an enormous accomplishment to humanity.'[24] (Schmidt's comments imply faintly that this access to information will of itself make people smarter, rather as if you could gain knowledge by simply sitting in a library. Experience tells us otherwise.) Amber Case, a 'cyborg anthropologist', told a 2010 TED Talk audience that 'We [now] have this thing called ambient intimacy. It's not that we're always connected to everybody, but at any time we can connect to anyone we want ... this is the first time in the entire history of humanity that we've connected in this way. And it's not that machines are taking over. It's that they're helping us to be more human, helping us to connect with each other.'[25]

The internet was now everywhere in our personal and professional lives, and that brought the good and bad of the internet into those liminal moments we hadn't previously noticed. Now your smartphone could fill those empty minutes. If you looked at your device, perhaps you could find something interesting – fascinating, even – taking place somewhere else. Perhaps it would be better than what was happening right in front of you. And since what you would be shown was a hand-picked window onto the world, that was almost certainly true.

But not everyone was comfortable with the way that people's behaviour was being funnelled through a few big networks. They 'force an architecture [upon the user] that allows for meaningful participation only if you play by rules that are designed for maximising profit, not optimum social and personal interaction,' observed the sociologist Zeynep Tufekci in February 2010.[26] Telling people not to use them, though, made as much sense as telling them not to use electricity, modern medicine, the phone: they had become woven into our lives. Rather like teenagers choosing to hang out in the corporate-controlled environment of a mall rather than at someone's house or the library or the park – because there's more room, no fussing adults, better weather – so internet users gravitated to a corporate social network rather than setting up their own blogs. 'Our social commons have moved online, and into corporate-controlled spaces,' she warned.

Even so, predictions about the positive power of the combination of smartphones and social media seemed to be vindicated when in early 2011 uprisings throughout the Middle East overthrew authoritarian regimes and dictators who had been in power for decades. 'We use Facebook to schedule the protests, Twitter to coordinate, and YouTube to tell the world,' tweeted the Egyptian activist Fawaz Rashed succinctly in March 2011.[27] Social media could route around state-controlled or censored mass media, such as when a video was posted on YouTube of the Tunisian government jet apparently being used by the president's wife for European shopping trips.[28] Ideas about democracy, and discontent at corruption, could spread far and wide. Egypt tried to prevent it by cutting off internet services and phone data, but the protests continued. The change happened fastest in the countries with the best phone networks and highest internet penetration.[29] 'The barricades today do not bristle with bayonets and rifles, but with phones,' observed Peter Beaumont, a *Guardian* reporter who was in the midst of the 2011 protests in Egypt,[30] where about 30 percent of the population was online, and more than eight million people were accessing the internet via their phone,[31] and nearly five million were on Facebook.[32] In the face of a certain amount of scepticism from people who, unlike him, had not been on the ground, Beaumont was certain that social media had played a role in the revolutions in country after country. Different platforms had played key roles in different places: sometimes Facebook, sometimes Twitter, sometimes YouTube.

All the rosy predictions seemed to have come true. 'Without social media, [Hosni] Mubarak's overthrow [in Egypt] would not have occurred,' wrote a group of self-styled 'cyberenthusiasts' in a 2013 analysis for the Rand Corporation.[33]

'Social media has created bridges . . . between activists, between even ordinary men, to speak out,' one Egyptian activist told the authors. What about the fact that social media use was a minority pursuit in those countries? The important fact, the Rand authors said, was that the organisers were using it: they could organise and reach more widely than they ever could have before. (They could also reach the highly connected, and highly dissatisfied, youth demographic.)

Biz Stone's 2009 musings about the open exchange having a positive global impact seemed to have been completely vindicated.

'I believe one of the most important outcomes has been the destruction of the old media regime and a move towards a new system, based on international standards of professionalism and objectivity,' wrote Jordan's former minister of state for communications in April 2011, looking forward to what he believed would follow. 'We will see the birth of a new credible and independent media.'[34]

Paul Mason, then a journalist for the BBC's *Newsnight* programme, tried to put the protests in context, observing that those behind them were graduates facing a future without employment, but who had access to social media: 'they can express themselves in a variety of situations ranging from parliamentary democracy to tyranny. Therefore truth moves faster than lies, and propaganda becomes flammable.' This 'oppositional youth', as he dubbed it, lived in 'a virtual undergrowth online and through digital comms networks.'[35]

Alec Ross, the senior adviser for innovation for Hillary Clinton, at the time the US secretary of state, said: 'If you are not open to social media spaces then you are not attuned to the dynamics on the street and you sacrifice both understanding and power.'[36] He added: 'We hope to maximise the benefits and minimise the negative impact of living in a hyper-networked world.'

Yet Facebook's executives seemed wary. 'We don't want to take too much credit,' said COO Sheryl Sandberg, speaking at the Paley Center in September 2011.[37] 'We did not march in the streets.' Facebook, she said, was an agent of change, 'giving individuals their identities, at scale, and the power of voice.' But what did Sandberg think lay in the future? She replied that the previous five years had been about proving that the idea of social networks built around real identity could work (though only Facebook had really done that; none of the other networks enforced it). 'The future is about realising what can happen when people share as themselves. This is about engagement, and giving people authentic voice, and what happens as a result. On that metric, we think we are at the very beginning.'

Zuckerberg was also cautious. 'My personal take on this is that social media's role is maybe a bit overblown in that,' he said in November 2011.[38] 'If people want change, they will find a way to get that change.'

Nobody questioned either whether in the future those spreading the viral messages might not be authentic. Nobody wondered whether that voice would be truthful, and whether the change being urged might come from outside the country.

And they didn't take much notice of the methods that the revolutionaries had had to use to get people motivated online in Tunisia and Egypt: one described it as 'boiling the internet', using everything from leaked documents to insults and obscenities to photos of someone urinating on a photo of the Tunisian president in order to motivate the youths who would have to take to the streets and confront the police, security services and even military.[39] Real revolution isn't done with mouse clicks. Facebook and Twitter had helped change the Arab Spring countries – but they had also unleashed a weapon that could be used at a smaller scale to destabilise legitimate governments. In a democratic country, might not the mobilisation of football supporters and activists to protest against authorities be just as easy to organise?

Newsnight's Mason, too, wondered about the assumptions some were making that the benefits only flowed one way, to those looking to free themselves. 'Are these methods replicable by their opponents? Clearly up to a point they are,' he observed. 'So the assumption in the global progressive movement that their values are aligned with that of the networked world may be wrong.'

And few people heard the rhetorical question put to Sandberg in the Paley Center, when she played down Facebook's role in the revolution: 'You're going to elect our presidents, build our brands and put us all back to work,' teased the interviewer, Pat Mitchell.

Sandberg demurred. But, she said, 'The biggest risk to our business is us. We'll get it wrong. We'll grow too fast. We'll grow too slow. The risk is us.'

By 2012, four in ten UK adults had a smartphone, and three-quarters of those aged under thirty-four said they'd made their lives easier.[40] Across both smartphones and PCs, half of adults had a social network account, and of that group two-thirds used them daily.[41] By 2015, the UK communications regulator Ofcom found that two-thirds of people had a smartphone, and used it for an average of almost two hours per day. They'd turned to that first rather than a PC.[42] The total time spent online had doubled in ten years. One in three adults checked their phones within five minutes of waking; for those aged between eighteen and twenty-four, the figure was half that.

The figures were similar in country after country: by the end of 2012, around half of all adults in countries including the US, Russia, Spain and the Czech Republic had social media accounts. Once you accounted for those who didn't use the internet, social media users were in the majority in twenty-one of the world's largest countries, except (inexplicably) Germany. People were becoming addicted to their smartphones.[43]

3

AMPLIFICATION AND ALGORITHMS: THE WATCHER BENEATH YOUR SCREEN

An algorithm is a fabulous name for something simple: a set of rules. We first encounter them as children, in those semi-magic tricks where you're told to think of a number, double it, add something to it, take away some other number (quite often 'the number you first thought of'), manipulate it a bit more and 'the answer is 2!' Follow the rules, and the outcome is certain.

Algorithms are everywhere in computer programming; knitted together, they produce programs of incredible complexity that can even be self-adjusting, tweaking themselves to account for changing imbalances between their input and the desired output – so in the above example you'd tell it the number you wanted to get, and the algorithm would change the equation to achieve that outcome. Or they can appear to mimic intelligence: if you train a 'machine learning' (ML) system – also known as a 'neural network', because it can behave like a collection of nerve cells in the brain – by showing it a set of 10,000 videos containing cats, and then another 10,000 that don't, it should be able to tell whether or not a previously unseen video features a cat.[1] At this point the ML system is not using anything as simple as a fixed recipe; instead it is more like a cook who searches for the perfect meal by constantly varying the ingredients in dishes and closely monitoring how eagerly diners eat them up. Although, to extend the analogy, the self-adjusting element means you can't look into the kitchen to find out how the meal was cooked; even for the creator of the system, all you know is what you get and what ingredients may have gone in.

ML systems are also improving rapidly. In 2013, the British company DeepMind set up a ML system connected to a number of video games such as *Pong* (two-dimensional table tennis) and *Breakout* (bouncing a ball off a paddle towards a wall of bricks). To master *Breakout*, the system was given control of the onscreen paddle, and simply told that a higher score was better, with no hints on rules, strategies or tactics. Within fifty hours, it could play *Breakout* and some of the other games better than the human assigned to the task. (When I visited the DeepMind offices not long after, a friend who worked there told me: 'Yes, we really do have a guy whose job is just to play Atari games.')

The algorithms improved exponentially. In 2015, a year after Google had bought DeepMind for its ML capabilities, a descendant of the *Breakout*-beating program called AlphaGo defeated the human world champion at Go, the Chinese game played on a 19x19 board, by 4–1 in a five-match series. It could legitimately be called the world champion. Until then, no Go-playing computer had ever beaten a top-ranked professional player in an even game, in marked contrast to chess, where the world champion had lost to a computer back in 1996. A couple of years after AlphaGo's win, another descendant of that program called AlphaZero taught itself to play chess and then beat the best dedicated chess computer in the course of a single day.

No surprise therefore that such ML systems are becoming more and more common, and increasingly companies hand over essential tasks to them. On Facebook, ML 'is the technology used to drive suggestions to Pages you might like or people you might know,' according to 'Annie L.', on the company's corporate careers page.[2] 'The News Feed team works on machine learning algorithms to surface the most relevant stories from people and pages in your connections. We do this by developing models that take in personalised signals from your social graph. I also love that Facebook recommends groups based on common interests,' she adds. (Facebook's habit of obscuring staff names so that they sound like encomiums for dubious skin treatments is peculiar, since a quick web search reveals Dr Annie Liu's full name, education and work background, including brief stints at IBM and Google.)

'Machine learning is increasingly at the core of everything we build at Twitter,' noted Jack Dorsey in June 2016.[3] 'It's powering much of the work we're doing to make it easier to create, share and discover the very best content,' he continued. His blogpost was announcing the acquisition of a London-based AI company called Magic Pony, which specialised in analysing video – a content form that machines struggle to analyse, unlike text. The analyst Rob Enderle suggested that being able to do that could turn Twitter into a real-time news aggregator if it could pull together disparate videos about the same topic: that, he said, would 'provide a level of coverage a [TV] network would envy'.[4] In turn, Enderle suggested, that 'could be both compelling and a huge ad revenue generator'.

But it might also be a means of exacerbating social warming: linking people to radical groups they might not otherwise have joined, encouraging virtual pile-ons, pushing people to the extremes of conspiracy theories – all because that would fit what the algorithms were programmed to measure as good, even if the humans affected wouldn't see it that way.

The trouble with algorithms is that you can design them to work in a way that seems sensible, or which you might even think is clever. But the interaction of two or more algorithms designed around the same purpose can create weird effects. Which brings us to Michael Eisen and the $23 million book.

Eisen is a professor of computational biology at the University of California, Berkeley, and one day in 2011 was alerted by one of his students to the price of a textbook on Amazon. Students often complain about textbook prices, but this was different. A fresh copy of an out-of-print 1992 textbook about fruit fly genetics was being offered for sale by one of the third-party sellers who use Amazon's platform as a virtual shopfront.[5] Other sellers were offering a used copy for about $35. But the seller of the virgin book had a more ambitious price: $1.7 million. (Plus $3.99 shipping. Clearly, they weren't going to leave money on the table.)

That would have been eye-popping enough. But another third-party seller was also offering an unused copy of the same book. Their price? $2.1 million. (Plus, yes, $3.99 shipping.)

Eisen knew there wasn't anything about the book that would justify such auction-house levels of pricing. So how had the price reached such a wild level? A glitch, he thought, or a joke.

Next day, he reloaded the page to see whether the prices had been corrected. Not at all: both prices had gone *up*. And over the coming days they kept going up. After a while, Eisen detected a subtle but deterministic pattern. Most of the time, bookseller A was exactly 27 percent more expensive than bookseller B. But once a day, B raised its price to within 0.7 percent of A's price. Later in the day, A would hitch its price upwards and re-establish the previous gap of 27 percent. The next day B would again edge up to within 0.7 percent of A's newer price. And so on.

But why? Eisen eventually decided it was an algorithmically driven sales tactic in the face of scarcity by two systems operating without any knowledge of each other.[6]

The rules they were using are simple to state. B's algorithm sought out the seller offering the most expensive copy of a book (or possibly the second cheapest; we can't be certain in this scenario) and set its price 0.7 percent below that. People doing a search would see the two and, bookseller B hoped, opt to buy the cheaper one.

A's algorithm, meanwhile, was built for what appeared to be a different business model. Bookseller A would fulfil orders for books by first buying them from other Amazon sellers, and then reselling them. A's algorithm therefore sought out the seller offering the cheapest copy of a book, and then advertised the same book with 27 percent added to the price, plus the same shipping charge. In a normal market, the price hike would provide enough margin to make money, and ensure no money lost on shipping. This was a viable method because A had a good 'seller reputation' on Amazon. As Eisen verified, A's reputation was better than that of B, the cheaper seller.

So how had the price gone mad? Because the market for the brand-new book about fruit flies consisted only of the two sellers, A and B, and their algorithms were too efficient. They were running every day, hiking their price without interruption because nothing was happening to stop them. Nobody was going to buy a book with a multimillion-dollar price tag. But the algorithms didn't know that. Like the program playing *Breakout*, they just had a task, not an understanding of its meaning.

By the time Eisen noticed it, the phoney price war had been going on for about six weeks. The algorithms continued chasing each other into the stratosphere: ten days after Eisen first looked, A's price hit $23,698,655.93. At that point some trigger was tripped: the next price for its non-existent book would have been over $30 million, which perhaps exceeded an internal Amazon price limit. At that point the prices reset, and B (which, remember, probably did have a fresh copy of the book that ultimately its retailer wanted to sell) brought its price down to $106.23.

Predictably enough, A set its price 27 percent higher, at $134.97. (In mid-2019, the prices for new copies ranged from $59.32 to $118.45. A year later, there were only two sellers, at $57.50 and

$66.47. None was the one Eisen had documented, though perhaps one of them had bought B's virgin book and was now re-offering it.)

That ability for algorithms with apparently sensible rules to spiral out of control when unsupervised is what scares many people who work closely with them. We keep giving bigger and bigger responsibilities to more and more complex algorithms, because the tasks that we're trying to manage are getting bigger and bigger. These days, algorithms – especially machine learning ones – are getting the plum jobs: playing chess and Go, spotting copyright infringement on YouTube, and increasingly choosing what we get to see on social networks.

What if the algorithms in charge acted beyond expectation like those bookselling ones – except that they had the power to alter the way people think, and how they perceive the world? And what if, on seeing that happen, the people who designed the algorithms declared that there was nothing wrong?

All social networks live by three rules. The first: get as many users as you can. The second: don't let go of the user's attention. The third, which is obeyed to varying degrees: monetise those attentive users as much as you can without losing their attention. If you do any one without the other two, you will have minimal success. Execute two well, and you might prosper. Do all three at once, and you can own the world.

The way to get as many users as possible is to make access free and fund it with adverts preying on attention. That's a price anyone can afford, because while incomes vary wildly, the 'attention economy' is the ultimate socialist paradise: everybody is allocated the same twenty-four hours per day. Time is the universe's most equitably distributed form of wealth.

Before Facebook, the model for advertising-driven online businesses was Google, where if you typed 'Sony camera' into its search box, you might be interested in buying one – so it made sense to show you adverts about Sony cameras, or other cameras, because you might be persuaded to change your mind. Google introduced an auction system for those ads, called AdWords, which took off and made lots of people, especially at Google, rich.

Facebook planned to do much the same, but because people wouldn't be searching for items they wanted to buy, people's desires would have to be inferred from a combination of things: their personal profile, the things they spent time looking at, the people they connected with. In time, Facebook would buy huge amounts of data about its users from credit-checking companies so that advertisers could target them better (an improvement for which the advertisers were happy to pay more). But the basic premise remained the same between Google, Facebook, Twitter, Instagram and later Snapchat: advertising keeps the lights on.

The big problem, though, was maintaining attention. Like TV networks in the past, every social network looked for the magic hit that would maximise the amount of time people spent with them. That entailed becoming as addictive as possible.

How though do you create an addictive experience from the random things that other people type into a status box asking 'What's up?' or 'What's on your mind?' At the time, people had to go and look at their Friends' 'walls' to see what had been written there. How could you instead bring the most interesting content to people?

That was the puzzle that a group at Facebook was chewing over ahead of the site opening up beyond college users in 2006. The prevailing method of showing content to users was the 'reverse chronological' series by then common in blogs: the top of the page showed the most recent item posted by your Friends, the second was the next most recent, and so on back through history.

This format is easy to generate from a database, as all that is needed is the timestamp of the post to output them in order. But it's vulnerable to overload: even if you only limit the potential posts to the past week, the average user with the average number of Friends could be shown any of 15,000 different pieces of content from Friends, Groups they belong to and Pages they've Liked.

Sturgeon's Law, which holds that '90 percent of everything is crap', applies particularly on social media posts, even from your friends. Zipf's Law – the 80–20 dynamic – also applies. In any large enough group, some people will post a lot more than others. If someone boring is churning out their fifteenth post of the day about potholes in the roads, you might not care, but they'll dominate the reverse chronological format. Pretty soon you'll grow bored, and move on to something else – perhaps Twitter?

Facebook realised that its busiest users could actually be the biggest challenge to growth. The problem resembled that of search in the years before Google, when search engines ranked sites based on how many times they used a particular word or phrase: the 'noisiest' pages would appear top in search. Only when Google began ranking sites using the 'PageRank' algorithm, which measured how pages pointed to each other as a measure of their perceived authority, did web search become effective.

The group at Facebook, which included Mark Zuckerberg, Aaron Sittig, Adam Mosseri and others, devised a system that revolutionised social media posts in a way comparable to Google and search ten years earlier. They called it 'EdgeRank' – a cheeky poke at the giant company – and created something that would wipe rivals off the web. EdgeRank was a system that considered each piece of content, which Facebook called an Object. The Object's 'Edges' were given values: your relationship to the poster (friend, family), the type of content (text, photo, video), how you had responded to an Object like that previously, how many interactions the Object already had, how old it was. Every Edge value also had a weighting that influenced how likely the system was to show that Object to you.

Just as Google's PageRank aimed to solve the problem of picking the site from the web's vastness that best answered any particular search, Facebook's EdgeRank aimed to solve the fundamental problem that as a social network grows, and connections between its members grow, the number of updates that could be shown to anyone tends towards infinity. Zuckerberg and his colleagues saw the EdgeRank system as so important that they filed for a patent, which was later granted, in March 2009.[7] It won't expire before December 2030.

Introducing News Feed, the product manager at the time, Ruchi Sanghvi, wrote that the features it would bring 'are not only different from anything we've had on Facebook before, but they're quite unlike anything you can find on the web.'[8]

There had been, and still are, plenty of successful sites not reliant on algorithms to choose what to show to people. Facebook led the way in abandoning any sort of human input to rank your likely interest in any particular piece of content. This, in turn, opened the door for all the effects of algorithmic amplification to take over.

News Feed achieved what its inventors wanted: people were able much more easily to find content they wanted. Yet its introduction in September 2006 was such a dramatic shift from what Facebook users had been used to seeing that many protested. Their Friends' activities were brought to them unasked; they felt it was intrusive compared to what had gone before. Now they were told if two people had just become Friends, that someone liked a TV show, or that someone else had just changed from being 'in a relationship' to 'single'. Many early users found it overwhelming. Danah Boyd, the sociologist, observed that the change meant 'Facebook is giving me the "gift" of infinite gossip. But I don't want it . . . Facebook says the News Feed is here to stay. This makes me sad. I understand why they want to provide it, I understand what users are tempted by it. But I also think that it is unhealthy [and] socially disruptive ... I also think that it will be gamed.'[9]

By 2011, Facebook had grown so vast that the simple rules that had been hard-coded in with EdgeRank weren't sufficient to satisfy the demands of holding hundreds of millions of users' attention; they were all too subtly different. It switched instead to a ML system that weighed as many as 100,000 different elements of an Object in deciding whether to show it to someone.

One important element in that decision was how much other people interacted with content, said Lars Backstrom, the engineering manager for News Feed Ranking, in a 2013 interview.[10] If a hundred users were shown an update and only a couple of them bothered to interact with it – commenting, Liking – 'we may not show it in your News Feed. But if a lot of people are interacting with it, we might decide to show it to you, too.'

Facebook even used Facebook internally to run the company: as it grew, communication between managers and staff was increasingly handled through a ring-fenced version (called Workplace) of the same system that ordinary users saw. In a hint of troubles to come, some employees learned how to game the system so their complaints or requests would be noticed by managers: express themselves more bluntly so that they were more likely to be distributed in the feed.[11]

The same system was used to create the advertising system. 'Micro-targeting' used the same functionality as the News Feed, except that advertisers were paying to reach people rather than posting and hoping people might see it. The advertiser could specify, say, that they wanted to be seen by men over fifty who lived in Pennsylvania and owned sports cars, and Facebook would show the ad to them. One important wrinkle was that the price of the advert depended not only on how many people it would be shown to, but also on how attractive Facebook's ML system thought the advert would be to that audience: adverts judged more interesting were less expensive, because they would maintain people's attention on the site. The game was all about attention, after all; boring content, even paying adverts, didn't fit the bill.

But the News Feed's contents can do more than just hold attention. In June 2014, the *Proceedings of the National Academy of Sciences of the United States of America* published a paper with a surprising title: 'Experimental evidence of massive-scale emotional contagion through social networks'.[12] Even more surprising was that the lead author, Adam Kramer, worked at Facebook and had run an experiment using nearly 700,000 of the site's users as unwitting lab rats to see whether their emotions could be affected by changing what was shown in their News Feed.

The work built on earlier social science experiments that had found that among personal groups, mood was effectively contagious: happy people could spread happiness; gloomy people could make you gloomy too. The basis for that was a twenty-year-long study on a non-electronic social network of more than four thousand people.

Kramer, with two co-authors from Cornell University, confirmed that Facebook statuses could do the same. They found that if you took away the happy posts from people's News Feeds, they responded with gloomier posts themselves. If you took away the glum posts, leaving happier ones, people cheered up and responded in kind.

The finding might seem obvious, but establishing it scientifically mattered, at least to Kramer and his co-authors, because if science is about anything, it's about demonstrating whether things that seem obvious are actually true. (Often, they aren't.) Proving that electronic social networks could generate the same effects as physical interaction was good news. Wasn't it? Furthermore, the work had been done on a far bigger scale, and much more quickly – the experimental phase only took a week – than previous studies.

However, the reaction from other scientists, and the public, suggested that this really wasn't seen as good news. First of all: where, exactly, had people consented to this manipulation of their feelings in a gigantic social sciences experiment? Second, who could or would rein in Facebook if it wanted to do other experiments? Sociologist Zeynep Tufekci, who is also an associate professor at the University of North Carolina, pointed out that this indicated much bigger potential problems under the surface. The existence of data and studies like this meant that corporations could model the behaviours and responses of individuals, not just narrow demographics. 'It is clear that the powerful have increasingly more ways to engineer the public, and this is true for Facebook, this is true for presidential campaigns, this is true for other large actors: big corporations and government,' she wrote in June 2014.[13]

Jim Sheridan, an MP in the UK parliament, told the *Guardian* newspaper, 'I am worried about the ability of Facebook and others to manipulate people's thoughts in politics or other areas. If people are being thought-controlled in this kind of way, there needs to be protection and they at least need to know about it.'[14] Tufekci agreed: 'I say we should care whether Facebook can manipulate emotions, or voting behaviour ... regardless of the merits or strength of finding of one study.'

After such a thorough roasting, Facebook rowed back hurriedly on the idea of running any more such studies. Kramer did publish a long defence of the work, saying he felt it was important 'to investigate the common worry that seeing friends post positive content leads to people feeling negative or left out.'[15] Facebook argued that people had consented to the studies by accepting the site's terms of use, which included the phrase 'We use the information we receive about you ... for internal operations, including troubleshooting, data analysis, testing, research and service improvement.'[16] Research! It said it right there, didn't it? But this wasn't just observation; it was manipulation, and in the scientific world that requires informed consent, which hadn't been properly acquired.

Facebook went quiet on the studies. But the News Feed algorithm had been revealed as a gigantic lever for anyone who wanted to manipulate people or their emotions for their own ends. Along with the micro-targeted advertising that made it possible to tailor adverts to tiny cohorts within much bigger groups, knowing that only they would see them wallpapered around or inside their News Feed, the site's news algorithms had turned into a sort of rat maze: once inside, you might become the subject of a giant experiment that you couldn't perceive from your vantage point.

The algorithms could shift your emotions as Facebook – or a savvy manipulator – wanted, and you wouldn't even be aware of what was going on. Subtle, smooth and analogue: social warming, quietly at work.

The problem overlooked by those who rely so heavily on algorithms, including ML systems, is that they are amoral – both in the sense that they do not know what morals are, and that they will act in ways that humans, aware of what we see as 'rules', will interpret as callous and transgressive.

The indifference to rules can surprise humans, who tend to see them implicitly. In 2018, an international group of researchers collected a set of anecdotes about what they called the 'surprising creativity' of artificial intelligence systems, and published them as a scientific paper.[17] Many of the examples were amusing, even delightful, in demonstrating how the programs had circumvented the rules the programmers thought they had set. One experiment tried to generate virtual creatures that could walk, measuring success by (virtual) distance covered in ten seconds. Rather than developing legs, as the experimenters had expected, the digital creatures evolved to be taller and taller and then fall over, thus winning despite not having taken a step; some worked out how to somersault in mid-air to extend the distance. Another, tasked with designing lenses for optical systems, refined and refined its product for accuracy. The lenses eventually offered, if made, would have been twenty metres – sixty-six feet – thick.

Some of the examples were more alarming in their implications for humans. One project sought to evolve an algorithm for controlling landing brakes for the arresting gear on aircraft carriers. An incoming plane trails a hook; on landing, that snags a steel cable pulled taut across the deck, which is attached to two giant drums on either side, with more cable spooled around them. As the cable pays out, brakes on the drums slow them down until the plane stops. The tricky part is applying just the right amount of braking force at the right time in a high-energy process lasting a few seconds and involving multi-tonne equipment. Computers, able to measure and respond to changes in microseconds, should be able to do so far better than a simple mechanical brake or one reliant on comparatively slow human reactions.

The algorithm's goal was to stop the plane, while minimising the force exerted on the combination of pilot, aircraft, hook and drums. The researcher, a Swede called Robert Feldt, set up a simulation in which the system would keep trying combinations of force and timing until it found the ideal solution, and left it running. To his surprise, the program claimed to have worked out a solution almost immediately, even for the heaviest aircraft. Feldt had expected the process to require multiple refinements. In fact, the evolutionary software algorithm's 'solution' had discovered a bug in the computer program that calculated the force exerted. If the algorithm applied a gigantic force on the braking system, the number wasn't stored – and so registered as zero. The aircraft and cable system would have been wrecked, and the pilot might have been dead, but the algorithm was satisfied that the job would have been done.

The authors noted that those working on self-teaching systems 'should be more sceptical of their ability to correctly specify robust fitness functions, and anticipate evolution iteratively, revealing such specification failures.' In other words: you're not so smart, and machines are relentless, and life – including the artificial form – will find a way to reach the goal you've set, but you don't know what rules it will ride roughshod over in the process. We tend to forget that machines don't have our human 'implicit' knowledge – the things that we know, and that we know others know, so we need not even mention them in a discussion. That's the problem with trusting ML systems whose workings we can't examine: we don't know what shortcuts they're taking or collateral damage they're ignoring in order to reach the goal they have been given.

Frequently, once such systems are unleashed on a social network, their creators find themselves wrestling haplessly against unintended consequences. 'Amplification is a natural outcome of the desire to get somebody to click on something, to convert somebody – whether it's to an ideology or clicking on BUY NOW,' says Aleks Krotoski, the presenter of the long-running BBC radio programme *The Digital Human*, who also holds a PhD in psychology. 'This is what we're being asked to do because it supports the financial model of the internet. But it also means that the human being is constructed in a way that is being reinforced. Reinforced and reinforced. And then hey presto, quelle surprise, ta-daa, here we are: isolated from one another and becoming more and more entrenched in our own views.'

Just how dangerous that entrenchment can be is illustrated by one of Facebook's biggest algorithmic missteps: the use of ML to recommend Groups and Pages to users. These are places where people can interact privately or publicly with content, with people who may or may not be Friends, and are one of Facebook's most important elements for keeping people on the site – always the overriding consideration in any design decision. The logic is that everyone's different: perhaps you're more interested than your spouse in the local running club; perhaps they're more interested in the local chess club. Perhaps you'd prefer to go to a closed Group to talk about team tactics for your upcoming football game. Facebook's ML systems could recognise the differences in each person's interests and recommend which Groups and Pages they should join. Those would all help engagement, and nothing satisfied the algorithm more than people spending more time on Facebook.

Unfortunately, the algorithm got a bit too good at its job. Certainly, it could find running clubs and chess groups. But as journalists and researchers pointed out, there are some groups of people who you don't want to form Groups: neo-Nazis, white supremacists, Islamic militants and jihadists, racists. The software didn't know that, though. Facebook's ML systems were excellent at spotting people who fitted the profile of other people who were already in such Groups. But the algorithms couldn't distinguish between a group that wanted to get together to go bowling, and a group that wanted to get together to beat people up. By using a combination of Groups that were public, 'closed' (discoverable by search, but entry controlled by an administrator) and 'secret' (undiscoverable, invite-only), extremists were able to make a comfortable home on Facebook. And the ML systems kept finding potential new members and suggesting, based on their 'interests' and online activity, that they might like to join those Groups – thus doing their recruitment, usually the hardest part of extremism, for them.

Facebook was warned about this repeatedly. In 2016, the Southern Poverty Law Center (SPLC), which monitors hate groups and extremists in the US, twice shared a list of more than two hundred Pages, people and Groups it felt should be removed from the site.[18] More than a year later almost all were still in place.

In fact, Facebook already knew that the Groups recommendation system was creating problems. An internal study in late 2016 had discovered extremist content in more than a third of the bigger German political Groups, where small groups of very active users had an undue influence on the position of the group. (This was a classic example of the 80–20 principle of Zipf's Law, and of another well-known effect by which closed groups with common precepts become more strongly aligned with those precepts – a sort of echo chamber effect.)

Worse was that the investigation found that nearly two-thirds of the people who joined extremist Groups on Facebook were urged to join by the company's own recommendation algorithm, based on what it saw as their 'interests' and how they matched with those who already belonged. Facebook was hothousing extremism by putting extremists in touch with each other – hardly a message that it wanted to get out when it was under scrutiny following three surprising electoral outcomes, in the Philippines, the Brexit referendum and the US, where Facebook content and advertising seemed to have played a significant part.

A study in Indonesia, which has the world's largest Muslim population, found in 2017 that young Muslims there who used the internet had more radical, less tolerant views compared to those who rarely went online. Interviews with convicted terrorists found that social media had a 'catalytic' effect: in nearly nine out of ten cases, the period between first exposure to radicalism and offence was less than a year. Before social media use became widespread, that lacuna would typically be between five and ten years. The social network was accelerating the problem, stressing the fault lines in society. That made it more important than ever that if Facebook did help create those Groups, then it should have moderation systems in place to spot them and move quickly to take them down. But the moderation couldn't keep up with the ML systems. The problem grew worse.

Despite all these well-flagged problems, in February 2017 Mark Zuckerberg wrote a two-thousand-word missive about how to solve a problem that he felt he had diagnosed in the world: not enough people in Groups. 'Building Global Community' emphasised how the company's next focus would be on 'developing the social infrastructure for community'. Borrowing somewhat from *Bowling Alone*, the iconic analysis by Robert Putnam published in 2000 about the breakdown in Americans' social fabric, Zuckerberg seemed to want everyone to find a Group to belong to; the more the better. Even if we didn't go bowling together, we could at least discuss it in a Facebook Group. In June 2017, he set a target of getting one billion people to join 'meaningful' Groups; to help that project, Group content was given more prominent placement in the News Feed.[19]

Meanwhile, a different part of Facebook worked on a solution to the extremist problem; inevitably, it involved ML systems. Now there were two algorithms at work on the site: one trying blindly to get extremists to join Groups; the other trying to remove those Groups.

In April 2018, Facebook announced a 'successful crackdown' against terror groups using the site: its detection technology could now block 99 percent of content from Isis, al-Qaida and affiliated groups before user reports, the company claimed.[20] The systems could even root out old content; half of what was identified and removed turned out to have been on the site for more than 970 days, or more than two and a half years. For newer content, the on-site period was much shorter, at a little over two days. 'We're under no illusion that the job is done or that the progress we have made is enough,' wrote Monika Bickert and Brian Fishman, then respectively vice-president of global policy management and global head of counterterrorism policy. Their blogpost also pointed out that extremists had been using the internet and forums since the 1980s; in that sense, they implied, Facebook was just another innocent receptacle for bad actors.

Independent research begged to differ: Facebook had not got rid of the bad actors, and was still actively helping them find each other.[21] In research carried out in the second half of 2018, well after the 'crackdown' blogpost, an anonymous whistle-blower pointed out, with examples, that Facebook was autogenerating terrorist content – and then helping anyone who professed approval for that content to find and network with others. In the battle of the ML systems, the automatic generation of content, plus the eager work of extremists and terrorists, was keeping ahead of the removal system. Examining the content produced by more than three thousand 'Friends' of terrorist groups, the whistle-blower found after five months that less than a third had been removed. Bickert and Fishman's claim about 99 percent looked threadbare.

'The AI [machine learning system to remove content] only targets two groups out of the dozens of designated terrorist organisations: ISIS and Al Qaeda,' the whistle-blower's write-up pointed out. 'Even then, it fails to catch most permutations of their names.' The problem was even worse for self-styled Nazis and white supremacist groups, which were able not only to operate freely (because the anti-terror ML system was only tuned to look for Isis and al-Qaida content) but would also be offered recruits based on Facebook's algorithms. People didn't have to already be al-Qaida members, or Nazis, or white supremacists; Facebook's algorithms would decide whether they matched the profile of the other people in the Groups and recommend those Groups to them. As a free recruitment method, it was hard to beat. Remember Dr Liu's pride in how Facebook's ML systems would recommend Groups based on common interests? That was how extremist groups saw their numbers boosted on Facebook. Yet to the algorithm, the process had the same moral weight as learning to play *Breakout*: people are the paddle, and Pages are the bricks, and more engagement is a higher score. For two algorithms, operating blindly without knowledge of each other, it was a never-ending race.

If engagement was lacking, the pro-Group system would generate 'Local Business' Pages for topics pulled in from Wikipedia, the vast (and ever-growing) encyclopaedia. For one person who listed their job experience as 'former sniper' for the Somalian terror group Al Shabaab, Facebook automatically generated a 'Local Business' Page, replete with the Isis logo, also pulled in from Wikipedia. (Al Shabaab has no flag, but had pledged allegiance to Isis.) It created a 'Local Business' Page for 'al-Qaida in the Arabian Peninsula', helpfully combining both English and Arabic versions.

Another Page, for a Syrian terrorist group called Hay'at Tahrir Al Sham (a militant group considered to be a covert branch of al-Qaida), had garnered nearly 4,500 Likes by February 2019. And just to help out, Facebook's ML systems then autogenerated 'Celebration' and 'Memories' videos for them, which they could share with Friends.

It wasn't just Muslim terrorists. The whistle-blower's research found thirty-one Pages and other content for white supremacist groups in the US, including one group that the US Department of Justice had identified for kidnapping and attempted murder. For the leader of one, the system autogenerated a business Page for his group – called Stamp Down Honky.

Lawsuits brought against Facebook in 2015 and 2016 for allowing such content to exist foundered because of the get-out clause of Section 230: it wasn't Facebook's fault that people posted such content, and it could remove what it wanted.[22] The topic that wasn't examined in court, though, was whether Facebook should face any liability for helping or encouraging such groups to organise.

'We do not allow hate groups on Facebook, overall,' Zuckerberg insisted to the US House Energy and Commerce Committee in April 2018.[23] 'So, if there's a group that their primary purpose, or a large part of what they do is spreading hate, we will ban them from the platform overall.'

The topic wouldn't go away, though. The very next month a counterterrorism researcher received a wave of Friend suggestions for extremists in the Philippines after he looked at some news about an Islamist insurgency there. 'Facebook, in their desire to connect as many people as possible, have inadvertently created a system which helps connect extremists and terrorists,' said the researcher, Robert Postings.[24] As he pointed out, people could be radicalised in as little as six months. The half-life of an Isis profile was longer than that, even with Facebook's systems. Furthermore, people who protested at being suspended could get reinstated.

Facebook responded to his findings by saying that '99 percent of ISIS and Al Qaeda-related content we remove is found by our automated systems.' This was the same statistic that Bickert and Fishman had quoted, though there was no objective way to evaluate it. 'But there is no easy technical fix to fight online extremism,' the company added.

The latter was true, yet might there be a way to pause the recommendation of potential recruits to extremists? If the price was that some normal people would have to look a little harder for a local running club's Facebook Group, or hunt through the profiles of Friends of Friends, would the world be worse off than it was by having a larger number of runners *and* terrorist recruits?

Nobody had the power to force Facebook to reckon with that question, however. To the algorithms, all that mattered was *more*: more Friends, more Likes, more Pages, more interaction.

To Alex Stamos, who between June 2015 and August 2018 was the chief security officer at Facebook, that wasn't too surprising. His role involved both defending the computer systems from attacks and preventing the products' misuse to cause harm – which included terrorists using the site to organise an attack. 'The world's best machine learning resembles a crowd of millions of pre-schoolers [under-fives],' he said in evidence to the US House of Representatives Committee on Homeland Security in June 2019.[25] 'There are certainly problems which a humongous group of children could be taught to solve, such as sorting a mountain of Skittles into five smaller mountains based on colour.' But the system's apparent intelligence wasn't additive. 'Adding more students . . . won't allow them to perform more complicated individual tasks. No number of small children could work together to build the Taj Mahal or explain the plot of *Ulysses*.' Nor, of course, to understand what terrorism was.

When it came to the prevention of radicalisation – the topic that he had been called to discuss – the problem was that its first steps didn't look like radicalisation. Instead, it began with 'frame crystallisation', which consisted of identifying and agreeing on who was to blame for a situation, and what should be done about it. But ML systems find it very hard to distinguish between a discussion about a burst water main and one about overthrowing a political party – the latter being the sort of 'frame crystallisation' that could lead, in time, to problems.

And anyway, nobody at Facebook seemed to be listening. In May 2019, the Associated Press wrote about the autogenerated Pages and 'celebratory' videos for Islamic fighters, and pointed out too how longstanding Pages set up by extremists got around the filters: by using text embedded in images.[26] (Facebook insisted that the search systems would find that, but couldn't explain why they had failed.) Four months later, the same research team found more autogenerated content for people who listed their schooling as 'School Terrorist Afghanistan' and 'University Master Bin Laden'.[27]

The algorithms certainly weren't listening. Just under a year later, in June 2020, Facebook announced that it would stop recommending people to join Groups pushing the 'Boogaloo' movement in the US.[28] Extremism researchers had been pointing out for months that their members were plotting violence. By the time many of the Groups were shut down, they had thousands of members. Their plans didn't involve going bowling together. They had their eyes more on inciting a second Civil War. Three members had plotted on Facebook to bring Molotov cocktails to a Black Lives Matter protest; another had recruited a Group member to drive a getaway car, and then shot and killed a federal security officer.[29]

You didn't, however, need to use ML systems to filter people off into murderous groups to cause problems. Twitter became an example of how blindly trying to increase 'engagement' could have negative effects too.

Twitter was comparatively slow to adopt ML systems for choosing content, as much as anything because its basic model – people write tweets, and you see them in reverse chronological order – worked so well. What analysts call its 'product-market fit' was nearly perfect: anyone who wanted a public forum where they could post short notes with a high likelihood of their being seen by their friends, or where they could (try to) contact famous people, or contact people without first being a 'Friend' – a bit like accosting people in the street – had just what they wanted. That made Twitter slow to adjust, and led Zuckerberg to describe the company contemptuously as 'a clown car that drove into a gold mine': he couldn't believe something so badly run could be so successful.[30]

Early in 2016, Twitter finally shifted to an algorithmic timeline, just as Facebook had so many years before. Instagram did the same, at almost the same time, pointing out that 'people miss on average 70 percent of their feeds.'[31] In future, 'The order of photos and videos in your feed will be based on the likelihood you'll be interested in the content, your relationship with the person posting and the timeliness of the post.' A year later, Instagram said people were interacting more with the content, leaving more Likes and comments.[32] Another win for the algorithms.

Twitter too benefited from the change: according to Deepak Rao, Twitter's product manager, 'every possible' measure of engagement and attention increased.[33]

But what sort of engagement and attention? For a long time, people had wondered why conversations on Twitter tended to degenerate so quickly. Was it because they were limited to 140 characters, corralling conversation into sentence-long thoughts? Even after that was lengthened to 280 characters in November 2017, the same seemed to happen. It was strange that a network where you'd tend to 'follow' people whose views you were likely to agree with could turn into one where you ended up having rows, often with people you'd never come across. Yet you could witness it, or experience it, again and again. Storms of rage could flow across Twitter with no obvious mechanism: people could come under attack without having had any significant retweeting, quote-tweeting or screenshotting. You could say something, and then find yourself fending off furious people you'd never encountered before and who didn't follow you. How?

If anything could be said to be a characteristic of Twitter's contribution to social warming, it was the way in which anyone could abruptly become a victim. 'Each day on Twitter there is one main character. The goal is to never be it,' explained 'Maple Cocaine', an ironic Twitter account that nevertheless captures the truth of the matter.[34] It didn't matter if you had a large or small following; you could find yourself suddenly thrust into a spotlight of pure hate, assailed by angry users you'd never interacted with before. Yet that hardly ever seemed to happen with nice topics:

people rarely experienced a storm of praise. Disdain and outrage were much more characteristic.

Andi McClure wasn't trying to find out how Twitter's algorithms can turn people into furious, hate-spewing keyboard warriors. But quite by accident she posted a single tweet that unravelled the answer.

McClure works as a senior software engineer at a video games start-up in Toronto and has been on Twitter since June 2011. That doesn't make her a fabulously early adopter (the site started in 2006), but certainly not a latecomer either. One day she noticed a couple of tweets go by, from people trying to gauge depth of feeling about a topic. 'I saw someone post something like "SOCIAL EXPERIMENT, don't read the replies before commenting" and then a question, and the question was something random about social niceties,' she told me later.

McClure thought it would be funny to take that to its logical conclusion: 'To just say "Social experiment: Do NOT read the replies to this tweet before you reply to it" and then cut off there, no questions.'

It was half a joke, half 'I wonder what will happen', she explained.

What happened was this: within an hour, about a hundred of her 21,200 followers replied, with banal or silly responses. Overnight, another hundred or so followers also replied in the same vein.

But then something changed: the replies kept coming. Now, though, they were from people whom McClure didn't know, people who didn't follow her, and – oddly – who had very few followers themselves, below perhaps two hundred or so. How had they seen the tweet to respond to it?

This was deeply puzzling, because when McClure investigated, there was no obvious link from her tweet to them. They didn't follow her, and didn't follow people who followed her, so were unlikely to have seen her original tweet retweeted into their timelines (the endlessly updating list of tweets from all the people they follow) by people who followed McClure.

McClure had noticed that her original tweet had had comparatively few retweets – even a day later, only eighteen people had pushed it on that way.

So where were they finding it?

Suddenly, she realised. 'Clearly what Twitter's doing is finding people with low-flow timelines [because they don't follow many people, and have few followers to engage with them] and filling space in with "high engagement" tweets – "things you're likely to reply to too".'

The algorithm was at work, desperately trying to drive up 'engagement'. Twitter's algorithm was guarding against the continual danger that people will give up, drift away, go to a different social network, never come back, which would mean it couldn't show them the adverts that pay for the servers, the staff, the buildings, the lunches, the plane rides. Without people seeing adverts, Twitter vanishes. And the only way to be sure that people will see the adverts that are fed into their timelines is to garnish that timeline with tweets from outside their network. Interesting tweets. Controversial tweets – which actually means tweets that are getting lots of replies from people all over the social network.

Twitter's algorithm tries to push those to users by showing 'while you were away' – a selection made by the algorithm of tweets to present when someone logs back in. They're based on the user's past interests, and also what other people seem to be finding interesting.

'From Twitter's perspective, "engagement" is good,' noted McClure afterwards. 'The more people comment on, quote RT, respond to etc a tweet, the more people are Using The Site and probably enjoying it. [But] the experience of having *made* a viral tweet is The Worst Fucking Thing.'

There seemed to be a threshold, she found, of about 100 replies: once a tweet has had that many responses, 'some algorithm at Twitter kicks in and goes "I need to show this to more people!" And shows it to MORE PEOPLE. People you've NEVER HEARD OF who have NEVER HEARD OF YOU.' For her tweet – a *tabula rasa*, a social experiment with no content – that didn't matter. But imagine, she said, if she'd said something political, or expressed an opinion about a boy band. 'At that point, Twitter surfacing the tweet to complete randos means *abuse*,' she observed. 'It means Quote RTs and spreading through communities who will ONLY engage with it negatively.' (Twitter declined to discuss its algorithm with me.)

There's a syllogism at work, McClure realised. Twitter has an algorithm designed so that tweets with many replies get even more replies; in other words, Twitter's algorithm automates and amplifies pile-ons, where people metaphorically yell abuse at someone they've never come across because of something they said.

Or as McClure put it more briefly: 'Twitter has an algorithm that creates harassment.'

But the harassment that people might suffer doesn't matter to Twitter as much as being able to show ads to those same people does. The engineers might like to think that they've created something comparable with AlphaGo, a design that could out-think humans and keep ahead of them. But instead, the algorithm was creating collateral damage that it had no way to measure and wasn't designed to solve. Fixing it would require someone outside to reach in and do a reset.

But who would mess with an algorithm that, by mid-2019, had turned Twitter profitable?

That same reluctance to override a system with demonstrable harmful effects because of its ability to generate profits also surfaced at YouTube. By 2012, people were spending a total of 100 million hours every day watching videos there. But that wasn't enough to make the site recoup its running costs. The problem, its engineers realised, was that for the previous seven years the number they had paid the most attention to when considering how their algorithm should rank a video was how many people had clicked the link to it. Whether they watched beyond the first couple of seconds didn't matter; everyone who saw the video begin to roll was counted as a 'view', even if they clicked away from it within seconds of the pre-roll advert finishing.[35] It was like a restaurant that measured its success only on how many people came through the door, not how many finished their meal. Someone else was paying the bill: YouTube, via the advertisers.

In August 2012, Eric Myerson, then YouTube's head of 'creator marketing communications' – the person who gave video makers the good and the bad news – announced a big change: YouTube's ranking system would now give extra weight to videos based on how long people spent watching them.[36] This used a new ML system that Google had been working on for some time. Among those on the developer team was Guillaume Chaslot, a young French programmer with expertise in neural networks. As part of his PhD, he had worked on a team whose Go program became the first to beat a human professional (albeit playing with a handicap advantage equivalent to a couple of pieces in chess).[37] Compared to beating a Go professional, getting people to watch videos may have seemed easy. On the face of it, the problems look similar: at every move in a game of Go there are hundreds of choices, each leading to hundreds more, quickly branching to billions of possibilities, from which you want to pick the single best option within a few seconds. Similarly, YouTube has hundreds of millions of videos to choose from; the challenge was to pick the best option within less than a second of someone searching on a topic.

But the system turned out to be a monster, and one that

YouTube was unwilling to slay because it did do exactly what was wanted: it got people to spend more time watching videos, and therefore it generated more money from advertising.

The problem was what economists would call its externalities: the effects on the rest of the world.

The team built a two-stage ML system. The first stage trawled through the millions of videos available to show next and, by balancing the user's history and current video context, slashed that down to just a few hundred potential candidates to be shown.[38] The next stage weighed them against what was known about each video, with the single biggest indicator of video 'quality' being: did other people who had looked at it watch to the end? That was then weighed against user details such as what they had watched before, which channels they subscribed to, what their Google search history looked like, where they lived, their gender and age, and what sort of device they were using.

By 2016, YouTube users spent more than a billion hours per day watching its videos, ten times more than in 2012, all under the watchful eye of the recommendation algorithm.[39] With a monthly total of 1.9 billion users, that meant every user was watching an average of half an hour of YouTube video every day. (For comparison, by the end of 2017 people were spending about 950 million hours per day on Facebook's News Feed, equivalent then to more than 40 minutes per day per user.)[40] But that 'average' figure hid an important point: quite a lot of those 1.9 billion users hardly ever visited the site. Some, though, spent much, much longer there – and those were the people that the algorithm tuned itself to please. By the recommendation system's logic, the videos that interested people who spent a lot of time on the site must be the videos that would also attract all those casual visitors: obviously, because those were the videos that people watched a lot.

What the engineers hadn't considered was what sort of people watch a lot of videos all the way to the end.

One day in 2016, Zeynep Tufekci, the sociologist, was browsing YouTube trying to check some quotes for an article about how Donald Trump appealed to his voters, and watched some videos of his presidential campaign rallies.[41]

YouTube didn't know that she was a social sciences academic writing an article. But the algorithm did know what people who watched Donald Trump rally videos usually watched afterwards, and what they watched after that. So the recommendation system got to work – aided by the 'autoplay' default, which would start the next recommended video seconds after the previous one. Tufekci watched in amazement as the recommendations moved to white supremacist rants and Holocaust denials.

Intrigued, she created a new YouTube account and started using it to watch Hillary Clinton and Bernie Sanders videos. The recommended videos soon included conspiracies about secret government agencies, and ones claiming the US government was behind the terror attacks of 11 September 2001. 'YouTube was recommending content that was more and more extreme than the mainstream political fare I had started with,' she noted. The same pattern was repeated with non-political topics: enquiring into vegetarianism morphed into advice about veganism. Looking at jogging videos led to marathons, which led to ultramarathons. Everything went to extremes. 'It seems as if you're never hardcore enough for YouTube's recommendation algorithm,' Tufekci observed, in an article headlined 'YouTube: the Great Radicalizer'.

The problem was that some viewers definitely followed YouTube's recommendations towards the extremes. These were the 'hyper-engaged' users. To the algorithm, they were the ideal person. But you or I would call them obsessive. What other sort of person spends hours and hours letting YouTube play videos to them?

At first, the complaints were about young men being radicalised by extremist Islamic preachers. But then journalists began discovering other sorts of radicalisation: videos made by white supremacists, even neo-Nazis, which would be recommended to people who looked at comparatively innocent content but would be led towards extremist views by an algorithm that didn't know what an extremist was – but which did like what they did online: watch lots of these videos.

By 2018, the problem had become embedded. That April two researchers published a study showing that there was a growing mass of right-wing YouTube content creators, and that the recommendation algorithm was acting as their gravitational field. 'While watching one of [conspiracy theorist] Alex Jones's videos won't radicalize you, subscribing to Alex Jones's channel and following the ever-more radical recommendations that YouTube throws your way might,' they wrote.[42] It could be even simpler than that: they demonstrated that having started at the YouTube page for the mainstream TED Talks, the 'recommendations' could take you to CNN, and then Fox News, and then to Alex Jones. They showed a similar but even more bizarre degeneration that started with music videos and ended at Holocaust denial. (Rather as Godwin's Law suggests that the longer any online conversation goes on, the more certain it is that a participant will be compared to Adolf Hitler, video radicalisation always seems to spiral towards disputing the reality of Nazi death camps.)

By the end of 2018, multiple shootings and killings had been blamed on radicalisation videos leading to obsessive people taking wild, murderous action. YouTube devoted significant resources to trying to stop people finding violent extremist propaganda.[43] Meanwhile, though, less obvious but equally dangerous ideas were spreading wildly, and people had realised that YouTube could pull people down a rabbit hole of conspiratorial nonsense – and that people who were liable to believe in one conspiracy theory were likely to take on others. 9/11 'truthers' who thought that the US government had somehow plotted the 2001 World Trade Center attack would be shown videos about the 'deep state' and global warming denial, or conspiracies about 'chemtrails', which are actually aircraft contrails. There was also a booming business in making such videos, because they could be monetised; unlike terrorism, conspiracy theories aren't illegal. And YouTube's ML systems would funnel anyone credulous towards them, sorting for the ones that kept people hooked for the longest. Persuasion about nonsense became a Darwinian survival of the fittest: the conspiracy videos that could capture the attention of the most people for the longest won out.

By 2017, Chaslot – who had left Google in 2013; he says he complained about the new algorithms' effects, while Google says he was fired over 'performance issues' – had begun to campaign on what he, like Tufekci, saw as a potentially dangerous system that was operating in the open, without any oversight.

Calling for transparency about the algorithm's effects, he carried out a study. If you searched 'Is global warming real?', about 25 percent of Google results and about 15 percent of YouTube results wrongly said it was a hoax. But for YouTube 'recommendations', the figure for a logged-in user was more like 70 percent.

Why the big difference? Because 'recommendations' were intended to keep you watching, not necessarily answer your question. It might not matter if a video had more 'dislikes' than 'likes' – that is, people effectively saying they didn't like the content – because the algorithm was all about keeping people staring at YouTube so they'd see ads. 'Having worked on YouTube's recommendation algorithm, I . . . came to the conclusion that [it] plays an active role in the propagation of false information,' Chaslot said.[44]

The other problem was the ratchet effect: once someone had gone down the rabbit hole, they were unlikely to climb out. Once you start believing a conspiracy theory, you'll be increasingly persuaded that everything outside it is untrue. On YouTube, you'll be shown more and more videos that seek to reinforce that view. Furthermore, the internet's ability to break down distance means like-minded people can find each other online, and bolster those beliefs in the face of reality. Before the internet, conspiracy theories were much harder to sustain because their adherents tended to be unable to find others who agreed with them. In the online world – and aided by an algorithm that rounds up the credulous like a sheepdog herding a flock – it's inevitable. The growth in belief in conspiracy theories is itself a form of social warming: an unwanted side effect of the system.

By making it possible for almost anything to be posted on YouTube, Google created the conditions where people distrustful of authority who also wanted an overarching explanation of all they felt was misaligned in the world could find it. Such thinking also tends to metastasise: people who believe that the Comet Ping Pong restaurant in New York was used for child abductions (the 'Pizzagate' conspiracy) would also believe that the children murdered by a gunman at Sandy Hook were 'crisis actors' or that vaccinations were a plot to control people. This is a common element of conspiracy theory believers, according to Professor Michael Barkun of Syracuse University in New York; in a 2016 paper he pointed out that this is because of their distrust of institutions, which interlocks with other conspiracy theories that might be about different topics – UFO cover-ups, miracle cancer cures, free energy systems suppressed by oil companies – and amplifies them.[45]

Measuring how long people spend on something, and picking content that rewards longer time spent, automatically leads to their being offered more content like that. And in YouTube, the AI has more content available to choose from than has ever been collected in one place before in human history.

The algorithm perceived hyper-engaged YouTube users as the best kind of viewer: long watch times, and perhaps plentiful comments. But it had no way of knowing whether they were also mentally ill, or being led down the path to terrorism, or an obsessive trying to push a conspiracy theory. Yet those people also shaped what everyone else was being recommended, because the algorithm wanted everyone to behave like them.

That's obviously dangerous. 'Engagement metrics create feedback loops,' Chaslot commented on Twitter. The logical, and visible, outcome was that 'eventually, hyper-engaged users drive the topics promoted by the AI,' he said. But, he pointed out, 'Some of our worst inclinations, such as misinformation, rumours, divisive content, generated hyper-engaged users, so they often get *favoured* by the AI.'

He called the resulting effect a 'death spiral' that was creating divisions in society. 'Is this "death spiral" good for engagement? Certainly: partisans are hyper-active users. Hence, they benefit from massive AI amplification. AIs were supposed to solve problems, but they appear to amplify others. What should we do?'

The YouTube engineers thought they were solving the problem of people producing videos with clickbait titles. But the algorithm they created saw the problem differently: that people weren't spending all their waking hours watching YouTube videos. From the system's point of view, perfection would only be achieved when everyone in the world was watching videos twenty-four hours a day. To the AI, the *Breakout* game isn't about demolition; to the book algorithms, absolute price doesn't matter; to the aircraft carrier system, the plane doesn't matter. What matters is what they've been told to achieve.

Alex Stamos, in his remarks to the Homeland Security Committee, pointed out that the two most important problems for social media companies to tackle were actually within their control: the ML systems powering their advertising, and content recommendation systems. This was 'partially due to the huge amplification [of visibility] either can provide, but also because these are the two components that put content *in front of people who did not ask to see it*.' It's that amplification by algorithm – suggesting Groups and Pages, showing tweets to interact with, showing you videos to capture your attention – that leads to the problems. But solving them would mean giving up user attention. And social networks just can't bring themselves to do that.

In January 2019, Google said it would change YouTube's recommendation system so that conspiracy videos – specifically, those which 'misinform users in harmful ways', notably flat-earth claims and 9/11 conspiracy theories – wouldn't be recommended so often.[46] It was the result of six months of work, though Google emphasised that the videos wouldn't be removed, just downrated. 'We think this change strikes a balance between maintaining a platform for free speech and living up to our responsibility for users,' said a blogpost from 'the YouTube team'. (The implication that until then free speech had trumped responsibility for users went unremarked upon.)

Chaslot was unimpressed. 'These are limited to specific types of harmful content, and go against the platform's business interest. Hence, the changes are likely going to be minimal,' he predicted. 'When I talked about these problems internally, some Googlers told me, "It's not our fault if users click on shit." But part of the reason why people click on this content is because they trust YouTube. The culprit is that users overly trust Google and YouTube.'

But, he says, 'recommendations can be *toxic*: they can gradually harm users in ways that are difficult to see without access to large-scale data. Researchers in universities around the world don't have the right data to understand the impact of these AIs on society.'[47]

Chaslot called for action by users, platforms and regulators: users should 'stop trusting Google and YouTube blindly. Their AI is working in your best interest only if you want to spend as much time as possible on the site. Otherwise, their AIs may work against you, to make you waste time or manipulate you.'

A March 2020 study by Chaslot and others at the University of California, Berkeley, found that for flat-earth and 9/11 conspiracies at least the change had worked. For many others, the results were equivocal at best.[48]

DeepMind had done its own research into recommendation systems late in 2018. The scientists wanted to know what might happen when the systems were let loose on data; one of the problems of machine learning is the way it can head off into mazes of 'learning' that to humans looks like rank misunderstanding, and create biased outputs. DeepMind didn't want its systems to fall into similar errors, and so a team looked at what happens when you let an AI do your recommendations.[49] They examined what happens to someone who is being recommended, and choosing, content from a set of potentially attractive choices that are effectively infinite. Even there, they decided, the recommendation engine 'can give rise to "echo chambers" and "filter bubbles", which can narrow a user's content exposure and ultimately shift their world view.'[50]

Ultimately shift their world view. It's a bland little phrase, but with colossal implications: an admission that YouTube's recommendation system can egg people on to change their minds based on what they're shown.

What then is the solution to the problems the algorithms create? Evan Williams, who created the Blogger platform and then went on to co-found (and briefly run) Twitter, described the drawbacks of algorithms trying to interpret human attention pithily in an interview in 2017. The problem that he observed is this: if you're driving down a road and there's a car crash ahead, everyone looks. It's irresistible. But online, the systems observing you interpret that to mean you want to see more car crashes, and will show you more of them, and will even *try to cause them*. After all, if you didn't want to see more car crashes, why did you pay attention to the previous ones?

That's certainly a problem social media companies wrestle with. One senior staffer (who didn't want to be named or have their company identified) told me that 'that's still a statement of the problem. And it's a very good way of looking at it, because half of the problem here is user behaviour, and half is systems and technologies trying to infer from that behaviour what the user wants, to show them something that will please and satisfy them. Essentially, you're taking the subconscious, the unconscious bias of humans and then linking them to a social graph and machine learning to feed those existing biases.'

Williams told the *New York Times* that 'If I learn that every time I drive down this road I'm going to see more and more crashes, I'm going to take a different road.' This seems, on the face of it, optimistic; he was denying the basic human behaviour he had just identified. Film-makers have known for years that action films can't have too many car crashes. The audience rewards them. The film-makers reciprocate. Replace the film-makers with machine learning systems that notice how many people come to the film, and you've got a feedback system that will give you car crashes all day, every day.

If YouTube's engineers had hoped that the problems with the recommendation system would be solved by the tweaks made in January 2019, they had yet another rude awakening a few months later. In June, a team at the Berkman Klein Center for Internet and

Society were investigating the impact of YouTube on Brazil's politics, where the year before the populist Jair Bolsonaro had been elected, partly through the impact of YouTube content. By accident, they discovered that some apparently innocent family videos of children in a swimming pool had thousands – no, *hundreds* of thousands – of views.[51] Delving deeper, they found hints in the comments that the skyrocketing views had come from paedophiles – and that the recommendation engine was unearthing similar videos for them. Even more shocking was that the problem was longstanding: YouTube had been alerted in November 2017 that the comments below apparently innocent videos about children included sexual remarks by paedophiles.[52] YouTube's response then was that comments would be turned off 'on videos of minors where we see these types of comments'.[53] As the Berkman Klein investigation showed, that didn't happen. But the recommendation engine dutifully worked away to give paedophiles gratification. Like bookselling algorithms chasing each other, like systems discovering a flaw that wiped out jet pilots landing on a carrier but scored well on simplicity, the programs didn't care or even know about the human cost. That was for others to discover.

4

OUTRAGE AND SCISSOR STATEMENTS: OUR TRIBAL MINDSET

The problems with social media can't all be blamed on the algorithms that observe us from beneath our screens. Social warming is about how people use these systems too. The systems that the social networks develop to spot what people are most engaged with, and feed back to them, have no concept of what sort of content is best going to capture people's attention. What will the algorithm feed on and amplify: pictures of kittens and lolloping dogs? Heart-warming stories of derring-do? Stories of stirring achievements? Or something darker?

Unfortunately, it's often 'darker'. Because sometimes the darkness is inside us. Social warming shows up, even without algorithms at work, when we enable emotions – especially negative ones – to travel faster and reach more people than they ever have before. The most dramatic example of non-algorithmic effects, as we'll see, is the messaging app WhatsApp in India. But first we have to understand why every time you peer into a social network, it seems to be suffused with anger.

Molly Crockett started her research career looking into the effects of serotonin, the chemical in the brain associated with happiness. One of her earliest published experiments in 2008 found that people negotiating over money were less likely to accept what they perceived as an 'unfair' offer (of a pay rise, for example) when their serotonin levels were lower.[1] To her delight, the work was picked up by the media – and then to her bemusement given headlines about cheese and chocolate, neither of which had figured in the paper. Why? Because the experimenters had given subjects a drink that reduced their serotonin levels by lowering their levels of the amino acid tryptophan. Some journalistic web-searching for tryptophan had established that it was found in cheese and chocolate. So there was the link: from cheese to more amiable decisions. Crockett was hardly impressed, but her work did at least bequeath the world the headline 'What a friend we have in cheeses'.[2] It was an object lesson in how grabbing attention matters more than accuracy.

Over time, Crockett's interests expanded to the psychological and neural mechanisms by which we make decisions. In July 2017, she moved to Yale University[3] to become a professor in psychology, where her laboratory web page has the punchy slogan: 'We study human morality in the lab and in the wild.'

That September she wrote a short commentary for the science journal *Nature*, titled 'Moral outrage in the digital age'.[4] It describes all the problems we tend to see online, but her analysis explains why they happen. 'Moral outrage' is the emotion you feel when you see something that breaks with your personal norms. (For that reason it's subjective, but you could start with 'a man punching an old, helpless woman'.) Outrage is an essential cleanser for a tribal society: someone who provokes that reaction from enough of the tribe will be cast out, putting them at serious risk from starvation or predators. Even though it leaves us upset, outrage draws boundaries around behaviour and also stabilises the group. Before the internet age, gossip told us who could be trusted in our personal social networks. But once electronic social networks could put us in touch with almost everyone alive, and with every form of human behaviour, the opportunities to trigger moral outrage exploded.

In normal life, violation of norms is comparatively rare. A 2014 study from Harvard University found that only one in twenty people in the US and Canada (from a sample of 1,252) reported experiencing or seeing 'moral or immoral' acts – behaviour that they felt violated social norms, such as unfairness, dishonesty, or (on the positive side) loyalty and self-discipline during a typical day.[5]

Yet even that figure of one in twenty seems, on reflection, a bit high. Certainly, you might expect that study participants would include someone driving their car through a red light, or failing to clear up after their dog, or (on the 'moral positive' side) helping an old lady with her shopping. But even 5 percent witnessing a full-on immoral (or moral) act feels like a big number.

Crockett thought so too, and looked more closely at the researchers' data. Five times a day, the participants were being asked to report via their smartphones what had happened. They also reported where they had encountered the moral or immoral acts: on traditional media, in person or online.

For every medium, immoral reports outnumbered moral ones by about two to one. But the key point Crockett extracted was that online reports dominated what people took note of. 'Immoral online' reports made up about 30 percent of the total, compared to 22 percent for in-person experiences, and less than 10 percent for traditional media.

Not only that: when the respondents gauged the strength of their reactions, they were far stronger to things they'd seen online than those they'd heard about or even seen in person.

This might seem counterintuitive: wouldn't you respond more strongly to something you experienced directly? Except if you're roaming the internet, you're far more likely to come across something shocking or surprising than if you're going about your everyday life. The world isn't like action films, with drama falling into our lives from the sky. On a global scale, however, almost everyone is carrying a camera attached to a network, and so the possibilities are endless. Not only that: the content doesn't even have to be novel. A post from a week, a month, a year, a decade ago can be brought back to life and recirculated to infuriate and inflame. Facebook's moderators frequently discover shocking content that has been removed multiple times from the network popping up for years afterwards.

What's more, being 'outraged' online is easy, and often consequence-free: if you call Donald Trump or Boris Johnson or Barney013598 an idiot online, are they really going to come after you? As Crockett noted, expressing your outrage effectively in person is comparatively difficult. 'Shaming a stranger on a deserted street is far riskier than joining a Twitter mob of thousands.' The car driver might come back and confront you; the dog-walker turn aggressive. She cited a real-world example from May 2017, when two men tried to intervene as a white supremacist shouted hate speech at Muslim women on a commuter train in Portland, Oregon. He stabbed and killed them.[6]

But you're never alone on a social network. Plus, as Crockett pointed out, you're 'not limited by location, time of day, or the probability of chance [physical] encounters with perpetrators'. Nor are we so worried about the effect on those perpetrators, because on social media in particular the people behind the accounts are effectively invisible: shaming someone represented by the pixels of an avatar has far less emotional cost than shaming a person standing in front of you.

So what happens? We're more likely, to share content that has an element of moral outrage; optionally, we can emphasise our purity by saying, 'Look at this, it's appalling!' as we do so. (René Girard might have liked the implicit mimetic and scapegoating behaviour of such actions.) A study in 2017 by a group at New York University examined more than half a million tweets about three topics that

Americans find polarising (gun control, same-sex marriage and climate change).[7] It found that using 'moral-emotional' words such as 'fight', 'greed', 'evil', 'shame' and 'fighting' increased how far a tweet would spread by 20 percent per word.

Even though we in the modern world have little need to eject people from our tribe, moral outrage provokes something deep in our psyche: the impulse to call it out can be irresistible.

The algorithms, meanwhile, don't know about morals, emotions or outrage; they simply notice what sort of things people like to share. Which means that if we share content that we find outrageous, the algorithms will respond by selecting more content we'll find outrageous, and serve that back to us, and we will respond by finding it outrageous and sharing it.

Renée DiResta, technical research manager at the Stanford Internet Observatory, describes the feeling: 'How often have you picked up your phone, scrolled through Twitter and thought, "There's no like fights in my feed, this is kind of a boring day"? We're wired for it now. It's not "Oh look, here's this interesting comment by this historian tweeting about a thing he knows, let me go and click in and learn more." It's "Wow, no one's fighting in my feed today, how is this possible? There's got to be a fight somewhere, I'm just not seeing it yet."' The reason why is very simple, though, she says: 'Everyone – media, social platforms, advertisers – is competing for that finite amount of your attention that they can get, and they have to do something that's going to grab you.'

Thus outrage, or its simulacrum, is both easier to feel, and reinforced online. Yet that brings drawbacks. Social punishment is a form of 'othering' – making someone seem outside the group, and even less human. Exacerbating outrage, the fuel of some of the bigger platforms, could thus increase social polarisation by making the targets of outrage seem less human. 'If moral outrage is a fire, is the internet like gasoline?' Crockett wondered.

A while before she posed the question, an answer of sorts had already been provided in India.

The police in Muzaffarnagar thought they'd done enough to stop the unrest. The city of about half a million people is less than 100 miles (145km) north-east of Delhi, in India's northern state of Uttar Pradesh. In a shocking incident, two Hindu boys had argued with a local Muslim youth, and then knifed and killed him. A number of Muslim inhabitants had retaliated brutally, beating up and lynching the boys.

Tempers seemed to be cooling, until a few days later a video surfaced on YouTube, claiming to show the two boys being beaten to death. The video was actually two years old and from Afghanistan, but people viewing it believed the footage was current. The day after the video began being shared on YouTube, the police blocked access to it – only to find someone had downloaded and begun sharing it on WhatsApp. Within a week, thousands of people in the area had watched it, and the area exploded into four days of violence between Hindus and Muslims that left more than sixty dead.[8] When the police arrested and questioned rioters, many had the video on their phone, and believed it was genuine. The inspector general of police in the area said his force hadn't realised so many people would have access to the internet on their phone, and that the ability of WhatsApp to spread the information so quickly had been a 'learning experience'.[9]

That was September 2013.

In the years following, India would repeatedly find WhatsApp at the centre of riots, lynchings and mob action. In June 2014, a 24-year-old Muslim IT professional was killed by an extremist Hindu mob furious at 'derogatory' images of their gods that had been put onto Facebook, and then spread on WhatsApp.[10] The murdered man was unconnected to the images or their circulation; he was walking home when the mob identified him as a Muslim.

When Facebook bought WhatsApp in 2014, it was already the most-used messaging app in India, Africa and Latin America.[11] 'It's very easy to use, and all of my friends are on it,' one twenty-year-old Indian woman told the market research company Jana at the time. Indians loved messaging huge groups of friends, family and colleagues: they would send cheery 'Good Morning!' images with embedded text each day.[12]

But the app also allowed the spread of fake news, and particularly of wild rumours about child-snatching and even organ-harvesting, which could flare up into violence. In 2017, as plummeting data costs meant that more and more people could get connected, the problem of deadly misinformation got worse. In May, seven men were lynched by a mob in Jharkhand: WhatsApp messages had been spread in townships warning that 'outsiders' were coming to steal children (a constant fear in India).

With India's literacy rate hovering around 75 percent, well below the world average, images tend to be more effective media than text. In one region, the concerns about children were determined to have been aroused by pictures of dead children – which, when the photo was examined more closely, in fact showed the aftermath of a chemical attack in Syria in 2013.

By summer 2018, WhatsApp had around 225 million users in India. Attacks on strangers got worse over the summer, and more than thirty people were killed, in separate incidents. In the wake of the continuing unrest and killings caused by viral misinformation, the government demanded that WhatsApp block such messages. WhatsApp responded by taking out newspaper ads and offering $50,000 grants to research projects that could identify ways to stop the spread of fake news on the platform.[13] It also stopped any message being forwarded more than five times, and added a 'Forwarded' label to re-sent messages[14] – a move that its engineers thought would demonstrate that a message wasn't vouched for.

Instead, the tweak had exactly the opposite of the desired effect: a BBC ethnographic study in late 2018 found that while people understood the label, they also saw it as an instruction to forward the message again.[15] 'The "Forwarded" tag does little to prevent sharing,' sighed the researchers. It was a stark illustration of the gap in understanding between computer-savvy engineers in California and less-educated populations grappling with the implicit assumptions of app interfaces. 'Forwarded' is factual, but what WhatsApp actually wanted to imply – and what its engineers understood, for reasons they probably couldn't have enunciated – was 'Unverified'. That distinction was lost on the millions of people receiving and passing on wild claims.

Given that sort of failing, we shouldn't be surprised that the pattern of WhatsApp being used to spread inflammatory content, triggering riots, being condemned by the Indian government, which then demands action, resulting in minimal (if any) measures, repeats again and again down the years.

In February 2020, a WhatsApp group with 125 members was used to spread inflammatory rhetoric about 'killing Muslims' and 'dumping their bodies in the sewer'; some offered to provide the guns needed to carry out violence. Then they turned to action: nine men decided to kill any Muslims they encountered, and for a period of about eight hours they killed or maimed any Muslim they encountered in north Delhi, according to charges filed in July.[16] The men would stop passers-by, demanding their proof of ID and that they chant 'Jai Shri Ram' – a Hindi expression that roughly equates to the Muslim cry of 'Allahu Akbar'. Nine Muslims were killed.[17]

Even though there's no algorithm at work, and this is purely down to humans' tendency to pass on outraging material, could WhatsApp prevent the spread of viral misinformation? Answering this question requires us to delve a little into how WhatsApp works.

A message written in WhatsApp is encrypted using a system called public key encryption that's best thought of as a key-and-padlock system. Your phone generates a single key, and an unlimited number of digital 'padlocks' that only your key can unlock. If someone wants to communicate with you, WhatsApp sends them one of your padlocks. When someone writes a message to you, that padlock on their phone is used to lock a digital box containing the message. The box is sent to you; your key unlocks the padlock. For group messages, the system is much the same: messages to the group are each secured with everyone's padlocks, and only decrypted at each individual's phone.

The upshot is that any message is encrypted all the time it is in its 'box', making the contents essentially inscrutable to everyone – including WhatsApp. The company can't peek at the content of messages, though they are stored in their encrypted (boxed) form briefly on its servers, where every message, image and video has a specific cryptographic 'shape' (formally, the 'hash'). If someone decides to forward something to another person or group, WhatsApp can see that the hash of the forwarding content is the same as content that is already on its server and send that rather than making the user upload it again. The user saves time and money (in data); WhatsApp can see how much, but not specifically what, content is being shared and re-shared.

Preventing content being re-shared then becomes a question of counting how many successive groups a particular hash has been sent to, and stopping it at a particular number. Cutting forwarding from 255 groups to just 5 in India (and 20 outside it) in 2018 reduced message forwarding by 25 percent globally, the company said later.

However, identifying any particular piece of content in order to remove it from the network isn't possible. To WhatsApp, that would be like trying to pick out a particular item in a giant supermarket while gazing through frosted glass. The Indian government demanded instead that the originators of messages should be identifiable. But that's a version of the same problem: identifying a piece of content and then tracing it back.[18]

Might the solution lie in educating people not to pass on misinformation? The BBC investigation asked WhatsApp users how and why they shared shocking content without checking whether it was true. The responses, the researchers noted, showed that 'Crucially, news is now considered to be as much about "how it makes me feel" as about "what it tells me".' Furthermore, anything that someone found important to them fulfilled the definition of news. (This is an important definition, as we'll see when we come to examine the impact of social warming on the media ecosystem: the definition of 'news' in the internet age is 'stuff I care about and/or stuff I want to pass on.') Social media had also lowered the barriers to entry for information into that 'news' space: absolutely *anything* could be 'news'. People thought standards of the traditional media had fallen – though that seemed to be more down to the explosion in availability of low-quality journalism, especially from satellite TV stations, lowering the average rather than a fall in the quality of the best.

India is a complex mix of identities, even within the Hindu religion, and tensions have been heightened in recent years by the rhetoric of the BJP Party's Narendra Modi, the prime minister. Notably, the more extremist members of groups push fake news – demonising Muslims, claiming the US space agency NASA uses Sanskrit invented by ancient Vedic sages as a computer language (it doesn't), and denigrating Indian media for 'ignoring' indigenous inventors. What the study found, though, was common to other countries: people were much more likely to believe wild claims that matched their political views, and disbelieve those which didn't. 'Ultimately, the main work that citizens are doing on messaging apps and social media is not rational, critical debate and discussion,' the BBC researchers concluded. 'They are effectively looking for validation of their belief systems.'

WhatsApp had its defenders: 'If we in India choose to use convenient messaging to form lynch mobs, that tells us more about India than it does about WhatsApp,' commented Mihir Sharma, writing for Bloomberg.[19] Indians, he remarked, had been killing each other long before WhatsApp came onto the scene: migrant labourers, nomads or those with learning disabilities had all been victims at some point.

That's true – but WhatsApp has raised the social temperature, and made it easier for those who are prepared to act dangerously to recruit others. No algorithms were needed; unfortunately, the users, acting on a deep-seated sense of outrage, provided all the amplification they needed. What was different was that the networks could reach more people more quickly and make the problem worse.

At the extreme, our subversive delight at outrage leads to complete polarisation, which can be focused around a simple-seeming statement. These are 'scissor statements'; their logical blades cut people into two groups, one on either side of the proposition, with no room for equivocation.

Arbitrarily splitting groups of people into two has a long and awful history. An early example in the Bible describes how the word 'shibboleth' – it means 'ear of corn' – was literally deployed as a weapon by the victorious army of Gilead against the defeated Ephraimite invaders. As the Ephraimites tried to cross the River Jordan to escape back to their territory, the Gileadites controlling the crossings would ask them to say the word. Ephraimites couldn't pronounce the initial 'sh-' sound, saying it as 's-' instead. It's a tiny difference in pronunciation, caused by a slight adjustment of the jaw.

According to the story, the Gileadites thus identified and killed thousands of Ephraimites.

Modern shibboleths exist – the twentieth century saw millions killed on the basis of their heritage – and have lost none of their power to divide. The newest ones are found on social networks, and they aren't single words any more. Now they're phrases, assertions, demands, and though their purpose is framed as creating a sort of unity, because they're framed as undeniably, self-evidently, incontrovertibly true, their effect instead is to divide the audience into two distinct groups with a mutual antipathy.

This is what 'scissor statements' do. (The phrase is an invention of Scott Alexander, a psychiatrist. He described them in a 2018 short story about a machine learning system taught to create such statements based on the most adversarial discussions on Reddit.) As the mechanism of outrage becomes more ingrained in social networks, a parallel evolution of language and behaviour has speeded up the amplification of viral emotions. Essentially, short-cuts develop to bootstrap their spread.

Scissor statements are often simple to frame: the United Kingdom should leave the European Union. Hillary Clinton would have been a better president of the United States. Trans women are women. Abortion at eight weeks is murder. But they're also impossible to prove. In scientific and logical terms, they're axioms – statements that you must treat as inherently either true or false. If you try to dig deeper into the logic, you'll hit a bedrock of belief that is summed up as 'because I said so!'

Sometimes conspiracy theories are used to create false dichotomies that work in the same way as scissor statements, even though they can be proven to be untrue: Barack Obama is a socialist Muslim, born in Africa and thus ineligible to be president. The 9/11 terror attacks were an inside job, using explosives to blow up the Twin Towers. Man never went to the Moon. The Earth is flat. Vaccines cause terrible illness and incurable conditions such as autism.

If you let them, scissor statements create tribes immediately, and with them the conditions for outrage: look at what the other tribe is saying about this phrase! Isn't it disgraceful!

As Alexander observed, 'Following the ancient imperative of evolution, if memes spread by outrage they adapt to become as outrage-inducing as possible.'[20] And social networks are the ideal environment for the rapid evolution of such statements; sometimes you can see it happen. '#Startanargumentinfourwords' said a hashtag on Twitter one day. In response, one user offered 'Hot dogs are sandwiches'. Another: 'Gun control doesn't work'. Yet another put forward 'Firearms are for militias'. Soon there were plenty more: 'Jaffa Cakes are biscuits', 'Jaffa Cakes are cakes' (a UK court was once asked to rule on this difference, for taxation purposes; because over time the confection gets harder, not softer, the court ruled that it is a cake, despite it being sold in biscuit-sized packets on shelves alongside biscuits); 'There's only two genders', 'Best president ever – seriously!', 'Most memes aren't funny', 'Women belong in kitchens', 'Evolution is a religion', 'Republicans are ALL racists', 'Hitler was a socialist'. There were many more. With a little more work, a little more selection pressure, a little more reward for going viral, the exercise might have turned up a few four-word phrases that could have sparked furious, prolonged debates. As those examples show, quite a few of them played on longstanding fissures in the social contract.

The logical endpoint of pulling people towards one or the other of those views is that they absolutely cannot straddle both; you have to be on one side or the other. Just keeping people interested isn't sufficient in a world of radicalisation; you want them to be on your side – or not, because having an opposition also validates your view. Imagine if nobody cared what flat-earthers thought, as was the case throughout the twentieth century.

The error that people make is to think that they can 'solve' scissor statements, and that they might therefore be able to convince people on the other side of the logical blades to change their minds. But scissor statements aren't intended to be bridged; they're intended to divide. They're a sort of perpetual motion of social networks, an inexhaustible fuel that never runs out: 'People are going to react to them, so they're going to get the engagement,' says DiResta. 'If people believe or disbelieve strongly in that statement, then they're going to react strongly to it – particularly things that they think are high-stakes, such as political conversations or that kind of thing.

'The way the ecosystem is constructed, when people react there is more "engagement". Because there's more engagement, it continues to be up at the top of the feed.'

Part of the problem is the device itself, DiResta suggests. 'The smartphone changes conversation in a number of ways. It lends itself to shorter, quicker types of communication – less nuanced. If you're battling for attention, you're not going to write a nuanced headline that encompasses an issue as fairly and balanced and neutrally as possible. You're going to write the headline or tweet or whatever that is going to communicate it as quickly as possible, and as sensationally as possible, so that you can compete with the rest of the attention-grabbing content that's out there.'

Bill Brady, a postdoctoral fellow at Yale who had worked with Crockett on a number of papers, suggests that to some extent this is the human condition. 'Our brains are naturally attuned to that type of thing, both in terms of seeing it in real life, but also seeing representations of it, and images and videos and so on from the media,' he says. 'Part of why moralised content spreads is because of natural tendencies of human psychology.' Outrage speaks to something deep in our psyche: the need for the tribe to stay alert to threats. Yet rather like our bodies' desire to stock up on fat in case we have to migrate across a land bridge during an ice age, our sensitivity to outrage may be a once-necessary characteristic that no longer works to our benefit.

What's more, we're now living in the moral outrage equivalent of the fast-food section of a supermarket. We're being encouraged to give in to these instincts so others can profit. 'What's unique about social media is there's someone else who's designing the environment that you interact in and you communicate in,' Brady says. 'And those people, those companies that design it – they don't necessarily have motivations in line with what you're trying to achieve when you communicate on a platform. I think a lot of people don't think about how your goals are always interacting, whether or not you appreciate it, with the goals of a company who designs the platform. We know, we have evidence, that the way you communicate actually is affected by these small design decisions.'

Outrage might not be good for your ongoing mental health – if there's always something to be angry at, when can you relax? – but it's certainly good business for the networks. (A *New York Times* article in May 2020, about the rows between Donald Trump and Twitter, described the model as 'the society-crushing pursuit of monetized rage'.)[21] And social networks need viral content. Remember the point about Zipf's Law: most people produce less than the 'average' amount of content, and don't have that many friends. Without virality, social networks would be caught in a strange trap, reliant on a small group who generate the most content – which the huge numbers of people not connected to them would never encounter. Broadcast media and newspapers would face the same problem if they relied on their audience to produce their content, but they don't: they rely on small groups of people who produce comparatively large amounts of content, which is then widely duplicated. Social networks avoid the duplication, but they're reliant too on comparatively small numbers of diligent users generating their content. What's the answer? A combination of suggestion (often called 'Explore', to give a taste of what the network is like, and suggested people to connect with) and virality. But what travels best is outrage, which splits people into tribes: the in-group and the out-group.

Another encouragement for performative outrage, as Crockett pointed out, is the absence of discouragement. Quite the opposite.

One element of the 'outrage system' that accidentally reinforces it is that we can't predict how much feedback we will get for expressing our outrage, nor when it will arrive. Unpredictability triggers compulsion; the same principle applies for one-armed bandits in casinos, where the uncertainty of whether you will be rewarded with a payout incites you to try again, and again, and just once more. When we log on to our social network, we don't know what we'll find; when we launch a little bit of content – outraged, humorous, factual – we can't know how it will be rewarded (or punished). It's a pull on the lever, a subliminal appeal to the gambler inside us. And the Likes, retweets, replies, comments, reactions to our input all form a sort of payment for our investment in cleansing the tribe, an equivalent of fur grooming among social apes.

Even before you consider the effect of algorithms, there are built-in reward systems in social networks that we respond to unconsciously: numbers of followers, friends who mutually follow us back or accept our 'Friend' request, 'Likes' and 'Favourites', retweet numbers, replies, comments. Each is a little boost to do more, to pull the lever once again. 'Have you ever noticed that Facebook's interface is full of numbers?' asks Ben Grosser, an artist and associate professor at the University of Illinois. In a video,he points out that any page on the site is 'an endless stream of metrics that quantify our user experience in Facebook'.[22] There's the number of new posts in Groups you subscribe to, the number of comments on a post, the number of Likes on a post, and so on. As a graduate student at the University of Illinois, he noticed that the numbers were the source of a subtle anxiety, leading him, for example, to judge the quality of a post by the number of Likes it received.

In response, he wrote a little program that he called the 'Facebook Demetricator' to remove all the numbers. His motive, he later told the art site Rhizome, was to get Facebook users 'to see how their experience is changed by [the numbers'] absence, to enable a network society that isn't dependent on quantification.'[23]

He also wondered how the visibility of those numbers served Facebook, which relies on a constant churn of new content to sate users: 'Where does it lead when quantity, not quality, is foremost?'

As he points out, Facebook controls which metrics are revealed, and since its goal is always to increase our engagement, those numbers are meant to tempt us to burrow down into *who* is behind that little red number, to find out who has Liked a particular post or comment, and perhaps follow suit. Similarly, when you're choosing whether to 'Friend' somebody, why does the site show you how many Mutual Friends you have? Does that matter? Is it the right basis for determining if you want to connect with someone? Grosser compares the experience of joining Facebook for the first time to walking into a gigantic cocktail party where at first you don't recognise anyone, and everyone is wearing a badge that says how many of the people in the room are their friends: 'Would anyone be content to keep that number at zero? At one, two or three? . . . You're encouraged to add another, to make that number go higher, to exceed in metric terms,' he told Rhizome.

The numbers push us towards doing things, even if we don't realise it, Grosser told me. 'We learn from a young age to do well numerically. We're taught you need to score highly, and get *more*. This is the message we have from day one. So, when you post a [Facebook] status about some esoteric topic that really matters to you, that you've spent a lot of time composing and thinking about, and it's really well crafted and it gets three Likes – you're dissatisfied. Then you post a funny cat video, and it gets 100 Likes. The system teaches you post more cat videos.'

Thousands of people downloaded his Demetricator browser extension, and he published a paper in a software studies journal on what the experiment had shown, titled 'What Do Metrics Want? How Quantification Prescribes Social Interaction on Facebook'.[24] The presence of those little numbers, he said, generates something akin to addiction: when we log in, we look at notification metrics to see how much attention we received while we were absent. During a session, users look at them repeatedly: 'the presence of a red and white number suggests metric success.'

At the time, people were posting more than three billion Likes and comments on Facebook every day. Though that sounds a lot, at the time there were around 1.3 billion monthly active users, and about 850 million daily active users checking into the site every day; that's an average of about seven Likes or comments per person every two days. If you apply Zipf's Law, and expect that 80 percent of the content comes from 20 percent of users, then roughly 170 million daily users were generating 2.4 billion of those Likes and comments – or about fourteen each per day – while the other 680 million were adding less than one per day, on average.

But that huge difference is invisible on Facebook. You don't get a figure telling you how many Likes and comments and posts the 'average' person generates, nor where you sit relative to them. You just get notifications implying that there are more Likes and more comments to be examined, especially relative to other people who have more of them.

Grosser also noted the excessive precision with which Facebook would tell you how recently new content had been published: '26 seconds ago', '23 minutes ago'; though this would become more vague as the event slid into the past: 'about an hour ago', and then more generally '2 hours ago' to include anything posted between 120 and 179 minutes ago. Such implied excitement about new posts 'present the news feed as a running conversation that you can't miss – if you leave for even a second, something important might pass you by,' Grosser commented.

A timestamp implicitly urges us to prefer the new over the old. Grosser's Demetricator for Facebook rejects that, and instead has two categories for time: 'recently' for anything within the past two days, and 'a while ago' for anything older. Some users found this unsettling, and one hacked the open source code to remove the feature and suggested Grosser incorporate it as a 'fix'. (He didn't.)

It's not just Facebook that uses numbers and signals to prompt us, he added: on Twitter, you can see the number of Likes and replies to a tweet change in real time, like a fruit machine reel spinning towards a payout. Naturally, Grosser also wrote a Twitter Demetricator, which removes the numbers of followers and Likes, and the other subliminal signals the site uses to grab users' attention. 'Twitter is often talked about as a very divisive space where polarisation of opinion is intensely present and foregrounded because the kinds of posts that can get great metric reaction are those that are highly charged, highly polarising, highly inflammatory. The presence of the metrics then encourages – or I would say *prescribes* – people's behaviour to be more polarising. To be more inflammatory, so that they can have the great metric reaction that gives them a very momentary blip of happiness.'

Danah Boyd, a sociologist at New York University, was a fan of the Demetricator: 'It's like an AdBlocker for stats junkies,' she wrote.[25] 'But why is it that it takes so much effort to remove the numbers? Why are those numbers so beneficial to society that everyone has them?'

The answer of course is that they aren't beneficial to society. They keep the pot bubbling for the social networks, constantly urging us to do more, to interact more, to get more approval, to trigger more outrage.

Sometimes, though, the network eggs us on too far, as in the case of Rhodri Philipps.

In July 2017, the fifty-year-old faced a judge, accused of sending 'malicious communications' – essentially, using the internet for threats – against Gina Miller, a mixed-race businesswoman who had campaigned successfully for the UK courts to rule on whether the government should consult parliament over the Brexit process.[26]

Brexit, by its nature, was a perfect scissor statement for almost everyone in the UK: the population was split into 'Leavers' and 'Remainers', and mutual outrage at the beliefs and actions of the other side became commonplace, as did personal abuse. And sometimes the outrage could turn to a form of anger.

In a Facebook post, Philipps had said he would offer £50,000 to 'the first person to "accidentally" run over this bloody troublesome first-generation immigrant'. What made the case notable was his status: Philipps was the fourth Viscount St Davids, a title that once (though no longer) came with a castle and a slice of land in Wales. Was that what one should now expect from viscounts – that they would direct racial abuse at people on the opposite side of a political argument, and even try to crowdfund their murder?

It turned out: yes. Philipps was sentenced to twelve weeks in jail for 'extreme racial abuse'. The judge was unimpressed by his claim to have belatedly realised his error after being criticised at an earlier hearing. 'This was a sudden conversion after many months when you have expressed racist views,' she said crisply. Philipps was also sentenced for another Facebook post about a father of eight children who was seeking a council home: 'I will open the bidding. 2,000 in cash for the first person to carve [the father] into pieces,' he had written. Philipps insisted the posts were neither publicly visible, nor menacing.

His case was a literal demonstration of something psychologists have long observed about the internet: people lose control of the ability to foresee consequences. That's because so often there aren't any. You write something rude that you would never say in person. The result: nothing happens. Outrage and anger can multiply.

Or at least it feels that way at first. But, in fact, there's a subtle process: it becomes easier to say rude things, because you don't see any consequences. The more rude things you write, the less anything that you'd expect to happen in the real world happens.

Empirically, it's obvious: it doesn't matter what you say. Outrage and anger are cheap. Until you're hauled into court for them.

In 2009, Chris Wetherell set out to create a new feature on Twitter, where he was a product designer. The feature was already a 'desire path' – something that people so wanted in the service that they were bodging a form of it themselves. (Real-world desire paths are the unofficial shortcuts that people create, such as by walking the diagonal across a square lawn. The solution is usually to pave where people want to walk.)

Wetherell built the desire path Twitter users wanted, but in retrospect he thinks that perhaps he paved a sort of road to hell.

He was in charge of building a neater version of the 'retweet', in which someone's tweet was copied and shown around to others. A retweet is a purely viral form of expression – simply amplifying what's already there – yet during Twitter's first three years, if someone tweeted a comment you thought worth re-sharing with your followers, doing so required an elaborate dance. First, copy the text into a new tweet, then prefix it with 'RT', and figure out the originator's Twitter handle and put that between the 'RT' and the copied tweet. Sometimes that wouldn't fit into the permitted 140 characters, requiring judicious editing. Once all that was done, you could press Tweet.

Despite the hassle, people persisted in doing it, and Twitter's engineers noticed. They had previously done the same for '@' to address specific people and '#' to organise discussion around a topic. The new retweet feature, introduced in 2009,[27] let users echo someone else's tweet with a single keystroke, meaning your timeline could suddenly contain tweets from people you didn't follow. In theory, a single tweet could be retweeted and seen across the entire network. Wetherell said later that there was little consideration of what sort of content was being shared by the primitive

'RT' method.[28] Biz Stone, the company co-founder, said that 'We hope interesting, newsworthy or even just plain funny information will spread quickly through the network making its way efficiently to the people who want or need to know.'

But enabling content to easily go viral created the potential for amplification of true and false information – and especially for outrage. In mid-2014, the 'Gamergate' controversy blew up on Twitter. Writing that October on the Deadspin website, the journalist Kyle Wagner described it as a hydra-headed mess of conspiracy theories and hate, a combination of 'incomprehensible Benghazi-type conspiracy theories about game developers and journalists', along with 'a somewhat narrower group of gamers who believe women should be punished for having sex', plus 'a small group of gamers conducting organised campaigns of stalking and harassment against women.'[29]

The standard under which they all marched, though, was that 'it's about ethics in games journalism,' which hardly anyone would have expected to be the topic most likely to spark a months-long online conflagration that would escalate from online abuse to physical harassment and death threats. 'It's a fascinating glimpse of the future of grievance politics as they will be carried out by people who grew up online,' wrote Wagner.

Gamergate turned into a battle where you could believe only one of two things: that the row was all about 'ethics in games journalism', or was not. If you believed it was, then everything arrayed against you was a sign of how justified your anger and your actions were – even if you defined the precise topic of 'ethics in games journalism' pretty widely to include not liking women, or not liking women talking about gaming, or not liking women objecting to men objecting to women (or women who talked about gaming), or just not liking the people who objected to you.

If you believed that it wasn't about ethics, in games journalism or elsewhere, then the opposition was an ever-shifting mess of constantly changing justifications for bad-faith arguments. Gamergate only ever involved perhaps 10,000 active participants, Wagner reckons. But though they might have been few, they were loud. 'Gamergate, in the main, comprises an assortment of agitators who sense which way the winds are blowing and feel left out.' More careful analysis of the Twitter accounts bore this out: looking at 340,000 accounts (some, or many, of which could have had a single controller), a research group from Greece and London found that they weren't angry so much as joyless. They had a higher proportion of 'friends' (followers who they followed back) than a random account, but they were nearly twice as likely to post something negative.[30]

The technologist Andy Baio analysed three days' worth of tweets as the battle raged that month. 'I've been using Twitter for eight years, but I've never seen behaviour quite like this,' he commented in his analysis.[31] He was most struck by the role of the retweet function. Of just over 316,000 tweets in three days around two key hashtags (#gamergate and #NotYourShield), about 69 percent were retweets. More than 38,000 accounts posted around the two hashtags, but that number halved once you removed retweets, which were being used as a bludgeon against conversation.

Twitter was caught essentially unprepared for such an inchoate yet determined incursion. As researchers at MIT and Michigan State University pointed out,[32] the same mechanism that had allowed participants in the Arab Spring to question their Middle Eastern governments now meant that anonymous users could target a small number of celebrity and media targets – who found themselves dragged, usually unwillingly, into a furious debate where they were outnumbered. But because Twitter had never considered the possibility of such an outcome, they had no means to tone down the discussion. Those embroiled could either leave the platform, or devote disproportionate amounts of time to blocking people. By insisting that it was 'the free speech wing of the free speech party', Twitter had left itself open to being used as the sexist abuse wing of the sexist abuse party.

Watching from the sidelines four years after he had left Twitter, Wetherell realised that the retweet feature had become a weapon without a countermeasure. An individual could not prevent hundreds or thousands of different accounts from retweeting someone else's comment aimed at you. Discourse became impossible because any thoughtful responses to you would be drowned out by the responses to those retweets from across the network.

The idea that that 'brigading' – acting in concert to quash other ideas, or harm someone's reputation by blaring untruths across the network – was enabled by the function he'd helped build concerned Wetherell. Gamergate, he told Buzzfeed News in 2019, had been a 'creeping horror story' as he realised that it was exposing the way in which people really wanted to behave; that this wasn't an aberration.

The collateral damage retweets can have on the quality of discourse chimes with an observation by Alexis Madrigal, a journalist at the *Atlantic*. In April 2018,[33] he wrote about how Twitter seemed to have changed since he joined in late 2007, when it had fewer than a million users.[34] 'Somewhere along the line, the whole system started to go haywire,' he wrote. 'Twitter began to feel frenetic, unhinged, and – all too often – angry.' His sense was that the anger principally emanated from retweets. While Twitter's settings let you turn off retweets by individuals, there's no single 'stop them all' setting; the average user would have to do it individually, which would be exhausting. But we live in the age of computers, so a programmer friend of Madrigal's wrote a script to do it. (There is also a clever search-and-mute solution within Twitter's interface that achieves the same end, but the average user would never find it.)[35]

The result, Madrigal found, was 'less punch-the-button outrage. Fewer mean screenshots of somebody saying precisely the wrong thing. Less repetition of big, big news . . . It's calmer. It's slower', he noted. Luca Hammer, who discovered the search-and-mute solution, agreed that his timeline was 'calmer' as a result. But after ten days, he turned retweets back on. 'I'm not convinced that it's a better experience overall,' he remarked.[36] Outrage is subtly addictive, and we miss not having it.

In April 2015, Twitter added another version of the retweet, called the 'quote tweet', which allowed comments 'on top' of the original tweet, so a layer of context could be added to the original tweet.

The idea of the 'QT', as it quickly became known, seemed sensible enough. But, in practice, it became a tool for ridicule and inducing 'dogpiling' on the person being quoted. 'The quote tweet is a uniquely agile enabler of anger, and that is where we run into a problem,' commented a writer for The Ringer in May 2018, remarking that its effect was to 'turn up the heat and the speed with which the storm travels'.[37] Whereas a retweet was often seen as an endorsement, either of its content or its author (compelling many people to supplement their Twitter biographies with 'RTs are not endorsements'), a quote tweet could make absolutely clear that you didn't endorse the contents. The effect often resembled someone walking out onto a balcony to an adoring crowd and announcing, 'You'll never guess what this idiot just said on the telephone! Let me read it back to you!' Someone with a large following could easily unleash a virtual pitchfork-wielding mob on the original tweeter. Jason Goldman, who had been Twitter's head of product when Wetherell devised the retweet function, told Buzzfeed News that the quote tweet was now the biggest problem on Twitter, because 'It's the dunk mechanism', used by the powerful against the weak to incite trouble.

When I contacted Wetherell, he told me that he was slightly more concerned about the retweet than the quote tweet. 'A potentially larger problem [than 'dunking'] is retweeting *without* quotes, which can be thoughtless, more easily automated, and provide less context,' he told me. 'Bare retweets can combine with context collapse [where the reference for the original tweet is lost] and confuse readers who come across tweets in hashtag searches, or as a result of Likes or retweets from people they follow, or as a part of tweets copied or embedded in other social media ... I think every time we re-share without context we risk orphan ideas being picked up by malicious parents.' That puts Wetherell, who initially made that process possible, in an uncomfortable position.

He added, of retweeting: 'That doesn't mean we shouldn't ever do it. But there's a power in amplification and I believe we each bear a really tiny but important responsibility in the aggregate.' Social warming is, after all, made up of millions of tiny decisions, each one contributing in its own way to the temperature of the debate.

Practicable solutions are hard to devise. 'I think re-sharing in all forms on social media might need enticements to think about consequences,' Wetherell told me. 'But how to craft those enticements isn't obvious.'

Some, such as David Rand, an associate professor at MIT, suggest that at least when it comes to false news, reintroducing some of the difficulty of the original 'RT' process could help. His studies have found that if people have to take even a little time to mull over what they're sharing, they'll still pass on true headlines, but are less likely to forward false news.[38] But would Twitter users really countenance returning to a world that most don't even recall, when they had to manually paste someone else's words into a new tweet? It would be like asking everyone in a suburb to give up their cars and use horses.

★ ★ ★

Early in 2017, Facebook began an internal investigation of how influential it might have been in the 2016 US presidential election and, to a lesser extent, the British Brexit referendum a few months before. Part of that involved the question of 'user engagement': how much time did people spend on the site, how much and in what ways did they interact with what sort of content?

The study found that most fake news and clickbait originated from small groups of extremely active, highly partisan users, most of whom were on the American political right. The effect of their content was to make Facebook users more politically polarised. But their heightened activity meant that they also had outsized influence, because Facebook's algorithms would amplify their apparently enthusiastic engagement.

A few months later, the research group led by the head of Facebook's News Feed Integrity team had enough data. 'Our algorithms exploit the human brain's attraction to divisiveness,' said a slide in a presentation to senior managers.[39] Without intervention, users would be fed 'more and more divisive content in an effort to gain user attention & increase time on the platform'. To fix that, the group suggested tweaking the recommendations and News Feed algorithms – though they warned that this might reduce engagement, and so be seen as 'anti-growth'.

According to a report in the *Wall Street Journal*, a number of senior managers, including Joel Kaplan, the company's policy chief, who had joined in 2011 (and previously was a senior staffer for the right-wing George W. Bush administration), rejected the suggestion. Anti-growth wasn't welcome. Partisan content and increased division were fine, as long as engagement went up. Facebook isn't especially interested in lowering the temperature of online discussion if that harms its business interests.

That's certainly how it looks from the outside, says Brady. 'It's tough from a scientific perspective, because we actually don't know how the algorithms work because they're not giving us access [to them]. But our work is highly consistent with the idea that these algorithms are producing more "outrage" content, more moral and emotional content in the networks, because we know that content is pushed more if it is engaged with more. And a lot of my work has shown that this type of moralised content tends to draw a high amount of engagement.'

Sometimes the engagement with outrage runs wildly out of control. In 2017, a Twitter storm blew up over an as-yet unpublished debut novel by an author writing for the 'young adult' category – a hugely competitive space, because a hit there can cross over to adult readers and spell success: the Harry Potter series and Hunger Games books first became popular as 'YA' fiction. Jesse Singal, a Brooklyn-based journalist who has documented (and been subjected to) numerous Twitter pile-ons – where a tweet or an author is held up for public shaming, which then seems to invite ever more intense metaphorical stone-throwing – watched in amazement. A single unfavourable pre-publication review on one blog catapulted the author Laurie Forest into a storm in which she was accused by the blogger of writing 'the most dangerous, offensive book I have ever read'. The accusation wasn't that the ideas espoused in the book were too radical, but that they seemed to confirm an unequal society split along racial lines – even though the plot was ostensibly about witches (and the book title was *The Black Witch*). Once the flywheel of outrage got going, people who hadn't read the 600-page book began weighing in. One accused the author of plagiarism; when Singal asked her to cite an example, she responded that a line had been lifted from *The Lord of The Rings*. Which was? 'Do not go where I cannot follow.' Seven words.

Rather like a Soviet court where your guilt was certain, and all that was needed was to establish your crimes, 'YA Twitter' collectively did not need to have read any of the book to know they hated it, and moreover that it should not be published.

In that case, the aftermath was more of a rebound. The book was published, the author toured cities, it reached the top of its Amazon category and Forest has written multiple sequels. When I revisited an article written at the time of the controversy three years later, I found that most of the outraged content the author had linked to had been deleted – including the original scathing review.[40] (The original blogger deleted her whole blog when she was named in a *New Yorker* article about YA's 'cancel culture' in late March 2019.[41] At the same time, she also demanded, on Twitter, that the magazine remove her name from the piece.[42] It didn't.)

Does that make criticism on Twitter a sort of aerogel, so insubstantial that it's lighter than air, and yet substantial enough that people must pay attention to it? The non-effect of trying to actively organise a political movement on social networks is remarkable: it's virtually impossible to find a movement that has been created entirely through one, with the possible exception of the French 'gilets jaunes' protests of 2019, which began through a series of Facebook posts.

Part of the problem is that social warming means everyone, particularly in tight-knit groups on Twitter, is on alert to break the outrage glass. As Wallace Sayre, a US political scientist, quipped about internal politics at universities, 'the reason they're so vicious is because the stakes are so small.' And certainly the sort of thing that can trigger a wave of outrage on Twitter in particular and then have real-world effects can look astonishingly trivial to someone outside the group where it blows up.

'Some random little thing that in the real world you wouldn't expect anyone to care about can go nuclear overnight,' Singal says. 'The speed of it is what's scary, because someone can just get ostracised so quickly, rumours can spread about them so quickly. It hypercharges the worst parts of human nature.' The topic *du jour* becomes a scissor statement: the book is bad and should not be published as it stands. (Being on the other side meant asserting that the book is fine for publication.) If the outrage is strong enough, a second-order effect takes over: some people will decide that they have even more insight into the topic's defects than the other people condemning it. This rapidly leads to what the journalist Gavin Haynes calls a 'purity spiral', where those who are insufficiently censorious are in turn censured, creating an ever-shrinking group who examine each other for sufficient purity in a process whose logical conclusion leaves just two people arguing over who can condemn something more than the other, and is thus more worthy.

But isn't the idea about social networks that they should, occasionally, bring out the good in us? 'I think it happens in communities that embrace certain values pertaining to communication and to civility,' Singal says. 'You can find good communities where they just don't let this bullshit take root. But the reason it hasn't happened on the broader internet is because all the incentives are to drive engagement, to elicit reactions that are as emotional as possible. If there was a profit motive that rewarded thoughtful, nuanced Twitter posts, that's what Twitter would look like. Twitter would encourage that. But the profit motives are pointed towards, you know, hysteria and cancellation and all that.'

For all their benefits in connecting us with our family, friends and the rest of the world, social networks are not without costs. We may be predisposed to take interest in lies, but social networks amplify that predisposition and make it commonplace, for profit. The Retweet button panders to our worst reflexes, and lets us amplify them effortlessly. As Brian Earp, a psychology researcher and philosopher working at both Oxford and Yale Universities, observed while watching a presentation by Brady, moral outrage spreads rapidly online (and particularly on Twitter) because we don't get the feedback we would get in real life, where part of the audience would probably roll their eyes in disapproval at what they saw as an overreaction.[43] However, 'there is no icon on Twitter that registers passive disapproval,' as Earp noted. There's nothing to temper the temptation to ratchet up the annoyance, the irritation, the anger, the full-on fury.

Earp wondered if there might be analytical methods to tamp down outrage. 'If Twitter collected data of people who read or clicked on a tweet, but did *not* like it or retweet it (nor go so far as write a contrary comment), and converted this into an emoji of a neutral (or some kind of mildly disapproving?) face, this might majorly tamp down on viral moral outrage' generated by a small subset, he suggested. Singal wondered whether the 'outrage train' might be slowed by the realisation that only some fractional percentage of the people who see a viral outrage tweet actually react to it.

The urgency that Twitter in particular can bring to our daily existence – with its proven ability to bring the first reports about topics that professionals care about – means that influential people will routinely spend hours there just in case something is happening, and as a way of sampling sentiment among those they care about. When things blow up, such as the row over a piece of unpublished young adult fiction, the different nature of the network takes over: 'It's a combination of the speed with which a discrete nugget of information or misinformation can spread and this idea of performing in front of an audience, and having these conversations take place in front of an audience,' Singal says. That has a perverse effect. The purpose no longer becomes to evaluate truth, or to try to find it; instead your visibility on the network means you have to signal to everyone how in tune you are with the values that you're apparently meant to hold. What's more, you have to respond quickly, because the panopticon-like nature of the network means that you can be condemned for not reacting quickly enough. Such condemnations are, ironically, almost always for not condemning someone or something else quickly enough. You'll hardly ever see someone being scolded for not congratulating someone quickly enough. Negative emotions – outrage! – nourish the feedback loop more strongly than positive ones.

Mike Monteiro, a long-time Twitter user who quit (for a time) at the beginning of 2018, observed in an article about why he left that the trouble with online outrage is that it's so indiscriminate.[44] As someone who suffered from depression on a recurring basis, for him one warning sign of a recurrence was being unable to distinguish between a big and a small problem, between something that could be ignored and something to be dealt with. And he saw life on Twitter as being the same: 'Every outrage was becoming the exact same size. Whether it was a US president declaring war on a foreign country, or an actor not wearing the proper shade of a designated colour to an awards ceremony. On Twitter those problems become exactly the same size. They're presented identically. They're just as big as one another.' The difference, he said, was that when that happens to him, he gets treatment for what he recognises as a problem; when it happens on Twitter, that is status A-OK, functioning as expected.

Continued exposure to outrage exerts a subtle psychic price. 'Doomscrolls' of content intended to infuriate and engage you can begin to feel overwhelming. The benefit of using Facebook, for example, might be better thought of in terms of the value that it brings to you. A subtle experiment in late 2018 managed accidentally to discover precisely how much people valued being free from the buzz of the infuriated, the imploring, the shocked.

A team made up of researchers from Stanford and New York University wanted to find out precisely what value – in hard money – people put on using Facebook, by offering to pay them to stay off it. They recruited 2,700 people in the US via adverts on Facebook itself (shown to 1.9 million people, of whom fewer than 2 percent clicked through).[45] They then asked each individually what they'd demand to be paid to deactivate their accounts, effectively putting them beyond use, for four weeks. The request came at a politically charged moment, covering the time before and just after the November 2018 US midterm elections – the first time the nation could respond at the ballot box to the presidency of Donald Trump.

Those who demanded less than a computer-generated (but undisclosed to them) number were paid and became part of the test ('Treatment') group; those who demanded more than the computer offered didn't get paid, and became part of the control group who continued using the site. Handily, about half of the total group demanded $100 or less, against the $102 the computer had been programmed to offer. (The researchers had expected that $100 would be popular: it's a nice round number that probably feels right to lots of people.) A significant number demanded substantially more: there were spikes at each multiple of $50, up to and above $500, a price that nearly 20 percent – perhaps optimistically – demanded. That pushed the average for the whole group to $180.

In the Treatment group, giving up Facebook freed up an average of an hour per day (in some cases, up to five hours). Rather than shifting to other social media, they watched TV or spent time with their friends and family. They also spent less time than before on news.

Significantly, the Treatment group also became less politically polarised – a measure that has otherwise been increasing in the US every year since 1984, with people who align with the two American political parties becoming further apart in their views about a range of issues. Not using Facebook led to significantly less polarisation on policy topics, and less exposure to polarising news. Yet the Treatment group were more likely to turn out to vote than the control group.

The most remarkable numbers were about happiness. Those who got off Facebook reported they felt happier, more satisfied and less anxious. (The positive effect was 25–40 percent as large as you'd expect if they had had therapy.) 'Active' users who had frequently commented on others' photos or posts reported being happier just as much as less busy users. But the lift in mood wasn't proportional to the amount that they used Facebook. That suggested that unhappy people used Facebook more, not that extra use made their mood worse.

The other surprise was in what the Treatment group said about their post-experiment plans, and what they actually did. After four weeks off the site, the participants said they planned to use it less – and, in the event, they did: a follow-up some weeks later found the Treatment group using Facebook about 22 percent less than the control group who had been on it all along. More than two months later, as 2019 began, 5 percent hadn't gone back at all.

The people who had taken part in the experiment seemed surprised by how much benefit they'd derived from it. 'I was way less stressed,' one said afterwards. 'I wasn't attached to my phone as much as I was before. And I found I didn't really care so much about things that were happening [online] because I was focused on my own life ... I felt more content.' Another had returned to playing the piano, 'which I used to do daily before the phone took over'. Yet another realised that 'It wasn't making me happy. I hate all of the interactions I had with people in comment sections.' They sounded a bit like ex-smokers who had rediscovered flavour in food and the pleasure of fresh-smelling clothes.

That wasn't uniform: some missed precisely that sort of interaction. 'I felt very cut off from people that I like,' said one. 'I'm kind of an introvert, so I use Facebook in a social aspect in a very big way,' said another.

When the four-week period ended, the Treatment group members were asked how much they would demand to be paid to continue staying off it. Now the numbers were much lower, by as much as 14 percent. Seen from outside, Facebook didn't appear so valuable.

Even so, the experiment seems to suggest that Facebook's 'consumer surplus' – the value that people reckon they derive from it, compared to the direct cost to them (which is zero) is gigantic. Extrapolating the data from the study to the 172 million Facebook users in the US would suggest that every month Americans derived $31 billion of value from the site. Over three months, a standard financial reporting quarter, that amounts to $92.9 billion in the US alone. During the same quarter that the experiment was carried out, Facebook's revenue in the US and Canada combined was $8.4 billion – implying that someone is getting a fantastic bargain, and it isn't Facebook, since it gets the smaller of the two numbers.

But there's a difference between giving up using something in return for payment and paying to use it. Mark Zuckerberg knows that well. Facebook could never charge $100 per month for its service, even though the experimental data might suggest that half of users would see that as a bargain: they wanted more than that to give it up, so don't they value it that highly?

Facebook's value, in this way, becomes greater the more people there are on it. If only you and 100 other people were users, your view of its value to you (and so the amount you'd pay to give it up) would be much lower: there would surely be other ways to contact those people. As the site proved when it was challenging MySpace back in 2006, the network effect creates a multiplier that makes a social network feel more valuable, even if we never use anything like its full capability. Our perception of the value of a social network lies in its potential, not our actual use of it.

But we tend not to factor in the cost of use, which is the calculation that was forced on the people who were paid to stay away.

That gave them a clearer idea of the effects. Some of the other ones – particularly the lowering of their political polarisation – weren't obvious even to the participants, but if extended to the whole of America's Facebook-using population would halt or begin to reverse a longstanding trend that has caused huge problems at every level of politics in the US. What's the cost of political paralysis? Of misinformation, or trolling? How much would people pay not to have those in their lives, or in the lives of those they are close to? There's no way to design an experiment to test that, because you can't create the different environments for people to live in, unless you move them to different countries and measure the change in their happiness.

This is what makes the impact of social warming so difficult to evaluate. In a world where we demand to know the value, the cost, the price of everything so that we can make economic yes–no decisions, the cost of people being more continually outraged, less happy and having an hour less per day to do other things is near-impossible to know. The same stymied decision-making is what makes action on climate change so impossible: how do you weigh the future cost of a planet in an unknown condition against the utility of being able to drive to the supermarket, or fly to a beautifully remote location on holiday, or cruise around an ecologically sensitive habitat? Just as we can see the benefits in *everyone else* not taking long-haul flights, we can see the benefits in everyone else giving up social media. But not us, not *quite* yet.

5

WORST-CASE SCENARIO: HOW FACEBOOK SENT MYANMAR HAYWIRE

If you had a map of south-east Asia in 2010 and were shading countries by the extent of their adoption of mobile phones – darker for more, lighter for less – you'd be colouring Singapore, Malaysia and Vietnam black: 100 percent or more, a phone for everyone.[1] The Philippines would be almost black, at 88 percent; Laos a bit lighter – 64 percent; Cambodia looking lighter still, at 57 percent. And so on until you come to Myanmar. It's the second-largest country in the area, after China to its north. Thailand and Laos are to the east, Bangladesh to the west, the uncaring Bengal sea to the south. The continental satellite picture shows a country washed by alluvial flows from the Himalayas, its south a network of river deltas. But in 2010 your pencil wouldn't have been troubled by Myanmar: at just 1 percent, mobile penetration that year was the lowest in the world, behind Eritrea and fractionally less than North Korea's 1.7 percent. Most phone owners were in the cities, where just one-third of the population lived.

And Facebook officially, and effectively, had zero users there.

What unfolded in Myanmar over the next few years shows the pervasive problem that social warming poses. As a country, it was as far as it's possible to be from the digitally saturated media-literate information economies where we would usually expect to see the signs of subtle warping of personal interactions by feedback loops amplifying outrage. Myanmar's experience showed that wherever social networks touch down, they will have the same effect, no matter the culture leading up to that point.

The British called the country Burma in the colonisation that followed a series of fierce, long wars in the nineteenth century. Independence finally came in January 1948, but didn't end the violence. Insurgent forces in various parts of the country troubled the nascent democracy, until in March 1962 a military coup led by General Ne Win created a Buddhist-Marxist junta that nationalised almost everything and ruthlessly suppressed opposition.

Today, Myanmar is a melting pot of ethnicities, wedged between different countries in a continent of wanderers. The government suggests there are as many as 135 separate ethnic groups in the population of more than fifty million, though activists suggest some 'ethnicities' exist only on paper, in order to subdivide and weaken minorities' voices politically, while others are ignored for the same reason.[2] There is a long history of tension between Buddhists, who make up 89 percent of the population, and smaller groups, particularly Christians (about 6 percent of all) and Muslims, who make up less than 5 percent of the total. The largest concentration of Muslims is the Rohingya, who mostly live in the north-western state of Rakhine, which borders Bangladesh; they make up about a third of the population there, with Buddhists comprising the remaining two-thirds.

Tension between the Buddhist majority and the Rohingya flared into broad conflict in 1978 and 1980, each time displacing roughly a quarter of a million Rohingya west into neighbouring Bangladesh. After that it had simmered, with the government refusing to give Rohingya full citizenship rights, to count them in the official census, or allow them freedom of movement, even between villages.[3] Sometimes their land would be confiscated, or they would be forced to build camps for the military, or drafted into the ranks.

Suspicion of the motives and legitimacy of the Rohingya runs deep among the Buddhist majority: many view them as illegal immigrants from Bangladesh, calling them 'Bengals' – a legacy of the colonial era. The junta after 1962 repeatedly implied that only Burmese Buddhists could be 'loyal citizens'. For decades, hate speech about the Rohingya has used wordplay that links them with 'rats' or 'fleas', and an oft-used rallying cry dates centuries back into the country's past: 'Race cannot be swallowed by the ground, but only by another race.' The implication is that different ethnicities don't vanish; they are overrun. Underlying many of the most divisive pronouncements is an irrational fear that the Muslim population will somehow grow so quickly that Buddhists will be outnumbered; wild claims that they 'breed ten times faster' than Buddhists are not uncommon. (That wasn't helped by a 2014 census that showed a Muslim population twice as large as the previous, longstanding, flawed estimates.)[4] One reason for Buddhist unease is that the country had been one of their religion's last redoubts when adherents were driven out of others.

In response, extremist elements among the Rohingya have demanded statehood for Rakhine, which it had until 1784. A couple of centuries hasn't eased the tension. Some have taken up arms against the military. Over the years, army reprisals against reported Rohingya incursions, and Buddhist attacks against Muslim civilians, became a periodic, depressing pattern, in which communications formed an important part. A 2016 study by Fortify Rights found that since 1938, such attacks had always been preceded by a rise in anti-Muslim messages being spread through the local population, whether in leaflets or pamphlets.[5] The severity of such ethnic conflicts was directly linked to the content of the propaganda and the ease of its spread. If spreading hateful content became easier, the outcome wouldn't be hard to predict.

The country didn't thrive through its dictatorship, particularly in 1987 when the ageing leader Ne Win superstitiously banned banknotes whose denomination wasn't divisible by nine, his lucky number. Millions of people who hadn't trusted the country's banks and had hoarded cash were abruptly rendered penniless. In 1989, a new military leadership declared the country would be renamed Myanmar, the formal version of the colloquial name the British had used. It also switched back to decimal banknotes, and away from communism.

However, in 2010 Myanmar was still the sick man of Asia. In 2008, Cyclone Nargis, one of the strongest ever recorded, had killed almost 140,000 people in the poverty-stricken Irrawaddy delta. Per-capita GDP, the measure of economic power per person, was the lowest in south-east Asia. The Asian Development Bank (ADB) and other foreign donors impressed on the junta that the outside world and the benefits of free trade could no longer be ignored. The first elections were held, and a quasi-civilian government – still overseen and picked by the military – was installed to begin a series of reforms. In 2011, media and internet censorship was eased, though various laws remained in place that would let the authorities arrest and jail people for offences involving content.

In a population of around fifty-two million, more than thirty million were farmers, and another five million did associated agricultural work, such as repairing farm equipment. The big industries were either agrarian, principally rice farming, or commodities such as gas, wood, and metals mining. Though unemployment was officially just 4 percent, a figure that hadn't changed for a decade, around a third of the population was underemployed, and a quarter lived in poverty.

And hardly anyone had a phone. The military regime hadn't wanted people to be able to contact each other and perhaps organise subversion or revolution. Phones, run by a state-controlled monopoly, were expensive, and hard to acquire; SIMs even more so, costing around $2,000 each, nearly double the average annual income.

The lack of connectivity and infrastructure, the predominantly agrarian lifestyle and economy, and the ethnic mix and history of tension, all made Myanmar probably the single place on Earth most unlike the technocratic, sophisticated and culturally homogeneous world of Facebook's Menlo Park headquarters south of San Francisco. But the latter's culture would soon decide much of how the former would see itself.

'Creating a harmonious society provides a foundation for inclusive and sustainable growth,' the ADB proclaimed in a 2012 report titled 'Myanmar in Transition', adding that 'The government can help create such a society by promoting understanding of the country's different cultures; by engaging in efforts at national reconciliation; by ensuring that members of ethnic groups have equal access to public services, jobs and other economic opportunities; and by building the infrastructure necessary for increased connectivity between rich and poor areas.'

The junta listened to the part about infrastructure. SIM prices were cut, and cheap mobile phones began arriving. Introducing mobile connectivity was expected to galvanise the economy, as it had in other countries. A 2012 report for the mobile network company Ericsson, by the consultancy Deloitte, on the potential economic impact of mobile communications in Myanmar suggested that just building out the mobile network would boost the country's minimal $60 billion GDP by between 3 and 5 percent.[6] That was even before accounting for secondary benefits to users from using the phones for tasks such as discovering market prices for rice, getting health advice, receiving emergency alerts and so on.[7]

At the time, mobile penetration had risen to about 4 percent – putting Myanmar ahead, at last, of North Korea and Eritrea, but still far behind its Asian peers.[8] Deloitte posited three scenarios for the further adoption of mobile phones: low take-up (up to 20 percent of the population having a mobile phone after three years), medium (up to 35 percent) or high (up to 50 percent). Previous studies have suggested that every 10 percent increase in mobile penetration leads to 1 percent in sustainable GDP growth.[9] On those forecasts, Myanmar could expect anywhere from one to four percentage points of long-term GDP growth. Deloitte also noted that mobile connectivity 'could also play an important role in enabling basic human rights, and in driving increased transparency in society'.

In 2012, the government-owned MPT network dropped the prices of SIMs to $200. At the start of 2013, they were cut to $150, and then in April to $2. Mobile purchases spiked.

In June 2013, after a competitive bid that surprised many for the absence of corruption, Norway's Telenor and Qatar Telecoms' Ooredoo won licences to operate mobile networks, starting from mid-2014. The world was on Myanmar's doorstep.

But there was a hidden problem that would only become clear as the outside world tried to interact with computer users inside the country working with the Burmese language, and which would make Myanmar even more isolated than years of underdevelopment and closure would imply. The written language uses an 'abugida', a writing system in which a single symbol represents consonants and secondary vowels; it's commonly used in south Asian languages. For example, 'myo', the word for city or township, consists of five different character elements that the writer must combine carefully into a written word so they don't overlap; otherwise the result will be illegible, or have a different meaning.

Like most languages, Burmese is not written using the Roman alphabet, and so using it on a computer needs a special version of the program that interprets keypresses and displays symbols. A longstanding computing standard, called Unicode, ensures that text written in any language can be passed between computers and interpreted correctly at either end. If someone writes an email to you in Russian Cyrillic script using a Russian keyboard, your computer set up to display content in the English language will still display the email in Cyrillic, because to the computer, each Cyrillic character is just a number. As long as it knows how to display that number, following the conventions of the Unicode standard, you'll see a faithful representation of the original on the screen.

Unicode allows for about 140,000 'numbers', covering hundreds of writing systems and symbols. For languages like Burmese, Unicode mandates sequences that the character numbers must be stored in to generate words, rather like the syllables of English words.

Many computing platforms, including the dominant desktop system Microsoft Windows, couldn't properly handle Unicode until 2005. Myanmar was anyway effectively cut off from the computing mainstream, so in 2005 a local developer eager to embrace the computer age produced a local system for generating onscreen text, called Myazedi. At $100 per user licence, it wasn't cheap, and within a year had been quietly pirated: an almost identical free version, called Zawgyi, appeared in 2006. With no effective competition from Unicode, Zawgyi quickly gained momentum among the PC users in the country, especially because it was used by Planet Myanmar, a home-grown news site that became an increasingly popular resource following protests in 2007 and the devastation from Cyclone Nargis in 2008.

But Zawgyi wasn't a computing panacea. It trampled over the Unicode system, using character numbers that were already allocated to other languages within the country.[10] Crucially, whereas Unicode only allowed the strings of numbers representing a word's glyphs to be stored in one way, so they would then be reconstructed correctly into the word, Zawgyi could store the text in multiple ways and then reconstruct the word. In effect, Unicode would always store 'confusing' as 'con fus ing', while Zawgyi might store it as 'ing fus con' or 'con ing fus'. So while Unicode only offered one way to order the character numbers for 'myo' to render correctly onscreen, Zawgyi offered ninety-six.

This seemingly minor technical difference had huge real-world implications. First, Unicode-only systems misinterpreted Zawgyi text and produced incomprehensible errors, and vice versa. Content written with Zawgyi can't be searched or sorted consistently – imagine trying to search a dictionary for a word when all the syllables are mixed up – and befuddles machine translation systems, because those learn how to translate between two languages by comparing large bodies of the same text in each. Zawgyi content is an inconsistent target.

By the time Facebook formally arrived in June 2013 with its site neatly translated into Unicode Burmese, Zawgyi was firmly embedded as the method for text entry and display by computers in the country. (Plenty of early adopters had already been using Facebook.) There was one upside: there was no chance of foreign influence on social media, because Zawgyi was such a localised product. Twitter's 140-character limit at the time wasn't friendly to written Burmese, either.

When Huawei and Samsung arrived to start selling smartphones in 2013, they looked at the most popular sites in the country and realised that if their phones didn't have Zawgyi pre-installed, they would lose out to any rival that did. So almost all the phones being used in Myanmar have Zawgyi. Even though Unicode is the 'proper' way to do it, everyone from smartphone makers to content producers, and particularly news websites, knew that in a competitive market, the benefits of sticking with what you have, even though inferior in some ways, outweighed the costs of defection to the 'superior' method, which could temporarily lose users and readers.

'The network effect is the only thing keeping Zawgyi alive,' remarked one local journalist in 2016, perhaps oblivious to the fact that 'only' the network effect keeps the telephone network, the internet or Facebook alive.[11] Zawgyi was as embedded as any of those, and constituted a barrier to external comprehension that Facebook in particular would struggle with for years.

However, people in Myanmar didn't mind that some of Facebook's text, in Unicode, looked odd on their own Zawgyi-based systems. A blank text box will accept input, because to the computer it's all just numbers, and those numbers will be reconstructed as the correct words by another computer, or smartphone, that uses Zawgyi. Facebook began to fill up with postings written in Zawgyi that would be displayed incorrectly to anyone using a Unicode-based system – though Facebook's systems could record, without understanding, what people were spending time on, what content they were 'engaging' with, what they were 'liking'. The Zawgyi users, meanwhile, understood the content perfectly.

Another facet of the arrival of Facebook in Myanmar wasn't obvious to many outsiders: the population didn't have years of digital experience that would allow them to approach Facebook – the app or its content – in the same way that more than a billion people did in the rest of the world.

Certain conventions built into Facebook's design are invisible, until you watch people who don't follow them. Here are a few: that you know what a 'network' is. That you want to be 'friends' initially with people you already know, and with others with whom you have lost touch. That you care about keeping your account personal, and about maintaining access to it (and so will respond to questions about it via email, the primary way of confirming an account's creation). That pressing the 'Like' button means you actually like – that is, approve of – the content you've just seen. That you will apply the same scepticism to content you see there as to what you see elsewhere. And, importantly, that there is someone who will oversee and enforce good behaviour on the network, and that you can reasonably appeal to them without risk to yourself.

Facebook assumes surprisingly large amounts of digital literacy because it began more than twenty years after personal computers became available, and ten years after internet access became broadly available in the West. In Myanmar, the world would discover what happens when you throw those assumptions away and give a population immediate access. It was the digital equivalent of giving everyone cars overnight, and not asking whether anybody had passed a driving test.

In January 2012, a Buddhist monk called Ashin Wirathu, then aged forty-four, was released from jail as part of an amnesty of political prisoners. Wirathu had been sentenced to twenty-five years' imprisonment in 2003 for using his sermons to incite the killing of Muslims, part of a years-long campaign he had run. On his release, he returned to stirring up hatred and racism. His words had previously been spread by pamphlets, CDs, video CDs and DVDs sold and distributed from market stalls. But now there was also the internet: blogging tools were free, and so was Facebook. He took to it with remarkable aplomb. Not only did the internet make distributing his sermons easier, it also brought in information from the world outside that could be used as a weapon. Though Isis was not yet world news – that would happen later in the year – the Islamic terror groups of al-Qaida and Al Shabaab were still making headlines, and that news had reached Myanmar as well. The idea of Muslim terrorists, incursion and conquest fitted perfectly with the story that Wirathu wanted to tell.

It didn't take long to have an effect.

In May 2012, the rape and murder of a Buddhist woman in Rakhine state triggered fighting and destruction. Over a week of violence, 200 people were killed, 115,000 were displaced and thousands of homes burned down. Wirathu's sermons were seen as having incited some of the early rioting; moreover, some observers worried that social media, meaning Facebook, was being used to incite hatred. In June, two Muslim Rohingya men were sentenced to death for the woman's killing; one other arrested with them was reported to have killed himself while in custody.

By this time, Aela Callan was already worried. An experienced Australian journalist, she had been one of the first to get a visa to live in Myanmar ahead of its political opening in 2012 and report on what was seen as the flowering of democracy. She saw the price of SIMs plummet. She saw censorship of online content relaxed – before, in fact, censorship of printed content was eased. And what really tempted people to buy a phone? The promise of the internet – or more precisely, Facebook. Mobile vendors quickly figured out how to create Facebook accounts by using disposable email accounts. 'People didn't have an understanding of email, so you would go to a shop, buy a cheap Huawei phone, and it would come preloaded with a Facebook account that wasn't attached to any email address or anything to do with you personally; it was just a burner from the shop,' Callan recalls. 'If you got logged out of your Facebook account, you'd just go back to the shop and get another account.'

Indigenous journalists were quick to realise the potential of the new devices and their connectivity, and among the first to adopt them. Facebook looked, to the eager eye, like the journalistic cornucopia: suddenly there were so many people who could be contacted so easily, and who were generating content all the time. All you had to do was look. 'That was the most usable app on the phone, and so everyone went on it. They thought Facebook was the whole internet,' says Callan. 'They didn't understand you could do Google search or that there was such a thing as email. They were using [Facebook] Messenger for their contacts. If they were writing a story, they would run their searches on Facebook.'

Ray Serrato, who worked on NGO projects in Myanmar from 2014 to 2018, noticed much the same thing: 'When I first started working in the country, the only way to contact people in a really reliable way was through Facebook.'

But Callan noticed that the excited local journalists, and others joining the social network revolution, didn't bring the scepticism to Facebook content that they had to the printed word. Facebook was seen as being mediated by the outside, trusted world. 'Everyone just assumed everything on Facebook was true. They saw that this wasn't a *New Light Of Myanmar* [the government-controlled propaganda newspaper] piece, this is from the outside world, so this must be true,' Callan says. When Wirathu's sermons began appearing on Facebook, journalists doing a search on 'Rohingya' would find his comments. 'So you've got Wirathu slamming Muslims, you've got all this anti-Muslim stuff that's coming out of the [United] States and other places, and people were saying, "Yes, this must be right," that there's a plot by Muslims to take over the country. And then journalists would write a story about that.'

On 20 March 2013, a fight in the trading town of Meiktila between a Muslim shopkeeper and a Buddhist customer – both female – led to a crowd forming, and then to a brawl involving around 200 people, which led to a rash of violent anti-Muslim attacks by a mob of around a thousand people. That left at least five dead and thirty-nine injured, and destroyed two mosques and an Islamic religious school.[12] Matt Schissler, an American PhD student who was living in the country at the time, described the uncertainty for those in nearby towns and cities who heard rumours that the trouble was spreading beyond Meiktila: government pronouncements were untrusted immediately, but could newspaper stories be relied on?[13] Or Facebook posts being shown off by van drivers?

Two days later, Eric Schmidt, then the executive chairman of Google, gave a speech in Yangon. 'Try to keep the government out of regulating the internet,' he told a group at a technical university.[14] 'The answer to bad speech is more speech. More communication. More voices.' He added that 'We have a chance to see how a new nation can shape itself ... and what I believe will be its extremely rapid social development.'

Callan, meanwhile, began filming a documentary, investigating the aftermath of the Meiktila attacks, and found that Facebook posts had been a key element in propagating some of the rumours inciting violence and denouncing Muslims. The documentary, released that September, called *Freedom From Hate*, investigated the eruption of violence between Buddhists and Muslims who had lived side by side for decades.[15] One of those who spoke on camera was Richard Horsey of the Crisis Group, an anti-conflict organisation. He commented that what had changed in the past couple of years was that 'information is readily available and transmissible, people are using Facebook, mobile phones ...' That spread might have sounded like a wonderful idea to utopian Silicon Valley types, but Horsey didn't think so: 'Every time you get an issue, every time you get a spark, it's much more likely to turn into a fire.'

Callan told me: 'The more I dived into it for the documentary, the more everyone was saying, "I read on Facebook that the Muslims were raping our women, I read on Facebook that the Muslims are trying to outbreed us, I read on Facebook that ..."

'And I was like – *hang on a sec!* Here's a platform that's designed for college kids in the US, and has this grand mission of connecting people. But what about a country like Myanmar, where connecting people who have never had any communications, like one side of the country to the other, with huge ethnic politics involved, is a bad thing?'

Serrato noted that while people were well versed in *how* to use Facebook, 'they certainly weren't well versed in assessing the credibility of information, or the kind of cues you might have on whether an account is legitimate: how often is the account posting, what are the links to the websites, where are they really going to?' About half of Burmese never began secondary education. They could read, but you'd question whether they were literate. And they certainly weren't digitally literate.

Later in 2013, Callan applied successfully for a fellowship at Stanford University, seeking funding to help educate journalists in Myanmar on how to recognise hate speech and how to shape narratives that didn't amplify it.

She was also concerned that Facebook should know about the trouble she saw brewing. The Stanford connection enabled her to get an interview on the morning of 6 November 2013 with Elliot Schrage, then in charge of Facebook's global communications, marketing and public policy; effectively, he was the company's No. 3, behind Mark Zuckerberg and Sheryl Sandberg. 'At the time, it was the height of Silicon Valley glory days,' Callan recalls. 'It was a huge privilege to get a meeting with the number three of Facebook. Everyone was saying, "Oh my God, you're so lucky, besides Sheryl Sandberg, he's the most important person."'

In the meeting, Callan expressed her serious concerns about the phrases being used to describe Muslims, and the potential for Facebook to be used to create harm in the fast-growing country. 'I told him that all of this had happened, and these were the problems, and the media's accelerating things, and people are using it for search.' The problem there was that searching on Facebook only shows content from inside Facebook, which would be inflammatory and would tend to pull people towards it: bigger and more active Groups tend to rank higher in Facebook searches.

Callan came away encouraged by Schrage's responses. In a follow-up email the same day, she expressed hope about the potential outcome. Schrage replied within hours, connecting her with a number of people inside Facebook and pointing out to them that Callan 'believes that Facebook is the path by which news is disseminated in that country' and that mobile access was due to explode. He also pointed out that she was focused on the problems of hate speech there.

There was a flurry of 'let's talk' emails. Then a week after the meeting, Vadim Lavrusik, one of the people Schrage had connected Callan to, effectively handed her over to the 'media partnerships team' – in effect, the press office. Callan realised her efforts were doomed: 'The PR team were very, very cautious with me because they were worried obviously that I would write about this, or make a film about this, and so they were very much managing me. They would be in on all the calls I had with any of the engineers.'

The problem, she discovered, was that the company had heard what she said about a country that thought Facebook was the internet and where three-quarters of the population would come online in the next three years. But the part about hate speech seemed to be a problem to be sorted some other time, by someone else. 'They even sent a PR team to Myanmar and asked me to help with the visas to basically do a video of "Look, we're bringing Facebook to Myanmar!" and I was, like, "Slow down, guys, you need to think about what's happening here, and what engineering things can be put in place that is going to make the architecture of your platform more viable for a country like Myanmar, that has a 'first do no harm' approach."'

Part of the problem was a serious mismatch between how Facebook's engineers believed users would understand the system, and how the users in Myanmar actually did.

The first was in the meaning of 'Like'. To Facebook's engineers, pressing the Like button was an indication of approval, singling one piece of content out from others. But Myanmar's users simply saw a box begging to be ticked. 'What happened was that people worked out pretty quickly that if they posted "I want to genocide all Muslims" in English – that's an actual example of something that was up there – it got removed. So they were posting it in Burmese. And then people would just go "Like". But they weren't using it as "Like". They were using it as "I've seen that". You'd watch them go through their feeds and they would just Like everything. It was just like "yes", like honking your horn – read, read, read.

'So then you end up with 4,000 Likes on "I want to genocide Muslims",' Callan recalls. At which point Facebook's algorithm would kick in and, despite being unable to decode the content, because it was encoded in Zawgyi, would note what it was programmed to interpret as enthusiastic 'engagement' from users, and thus promote the post to more and more people, who would read it and thus Like it, reinforcing the cycle.

The second problem was the cultural mismatch around the idea of 'reporting content', which Facebook relied on heavily to keep postings in line with its 'Community Guidelines', which no normal person reads. Searching Zawgyi-encoded text was a barrier to automated systems, so questionable content had to be flagged by users, who when they observed something objectionable were expected to pick out a drop-down menu on the offending post and choose an option labelled 'report violation of Community Standards'.

Those, though, are very loaded words in a country emerging from decades of repression and censorship. Callan recalls Facebook's engineers telling her confidently that people could just report offensive content, rather than Like it. 'Do you know what that *means* in a former military dictatorship?' she asked them. 'That means the police are going to come to your house in the middle of the night. Nobody wants to visit that on anybody.' The people of Myanmar didn't know what 'report' meant in the context of

Facebook content – who would review it, what the punishment might be.

The biggest cultural mismatch, though, was embedded in Facebook's ethos and its business model. To Facebook, preventing the spread of content is anathema. The system's whole intent is to connect people; if they can't pass content to each other, what's the point? Its rules for moderating content principally cleave to the American ideal of its Constitution – a Puritan ethic sprinkled with libertarianism, which allows guns and violence but bans genitalia, sex or nipples. This has led to the slightly perverse logic that the display of devices that can cause death and injury is fine, but that natural things that can create or nurture life are dangerous.

When it comes to what's categorised as 'hate speech', the rules are loose. 'General expressions of anger' are OK, but 'specific calls for a named individual to be harmed' are not; nor is 'any credible threat of violence'. Those will, in theory, be removed.

The process isn't automatic, though. A human moderator hired by (but not directly employed by) the company has to be alerted to the existence of the content, and then weigh it against Facebook's confidential rules. A computer could search English postings for words like 'genocide' and 'Muslim', and pass that on to a moderator. But a computer couldn't search Zawgyi postings effectively; that needed human beings who could read the language natively.

So for the most part, as Myanmar's population was getting its first taste of the internet, the responsibility for keeping the lid on discourse rested on the shoulders of Facebook's moderators. Or, as it transpired, moderator.

In 2013, thousands of posts began appearing on Facebook calling Rohingya 'kalar' (an ethnic slur used by some Burmese against darker-skinned groups) and suggesting, 'We must fight them the way Hitler did the Jews.' They were dogs, rats, deserved to be killed, the posts said. (As multiple organisations have pointed out, dehumanising language that equates groups of people with animals is a standard rhetorical method for inciting violence and even genocide.) Sometimes, the posts pulled in memes from far afield. Ray Serrato recalls reeling at a Facebook Page in Wirathu's Ma Ba Tha (which translates as 'Patriotic Association of Myanmar') group: 'I saw a meme that was referencing London [in the UK] becoming "Londonistan"' – an exaggerated suggestion that Muslims were taking over the capital – 'and the caption was also about "this is going to happen in Myanmar", and I thought this was *really* . . . ' He laughs, unable to find the words to capture his horrified amazement.

However, as Schissler noted shortly afterwards in a scholarly book about Muslim Buddhist relations in Myanmar, the arrival of such foreign tropes was important for those who wanted a narrative of potential conflict.[16] While Muslims might be a tiny minority in Myanmar, the subliminal message went, look how many there were elsewhere in the world! Look how they were taking over! 'There is a perception that Muslim power, global Muslim power, is very great,' Daw Aung San Suu Kyi said in a BBC interview in October 2013. Whether she was distilling or just observing Buddhist tension about Muslims, her words demonstrated that it was only growing. And, noted Schissler, Myanmar's population was now being exposed through Facebook to inflammatory videos and rousing slogans, the political form that – as decades of US political attack ads have shown – can be the most persuasive form of messaging.

Mobile phones were also becoming part of life. A survey by the International Republican Institute (IRI), carried out between December 2013 and February 2014, found that 40 percent of households had a mobile phone, and that 5 percent of the population accessed the internet at least once a week: just over half did so via a phone, 29 percent at home, 7 percent in cafes, and 5 percent at work.[17]

★ ★ ★

By 2014, Facebook's user base in the country had grown to about two million, or just over 3 percent of the population, and around half of people had a mobile phone.[18] In July, the messaging app Viber, which ties user IDs to a phone number, said its user base had grown to five million, from two million in February. Smartphone ownership and data use was exploding.

But so were problems. A riot was triggered that March in the southern coastal town of Sittwe by rumours spread on Facebook that a woman from an NGO had dishonoured a Buddhist flag.[19] Soon mobs formed with knives and stones, prompting the NGO staff to flee. An eleven-year-old girl was among twenty-one who died in the frenzy. Matt Schissler later told Reuters that over the next nine months he held a number of talks with Facebook officials about the extent to which the platform was being used to spread inflammatory rumours and hate speech. One example he cited was a post about an aid worker, consisting of a photo of the person and commentary calling him a traitor to the nation. Yet when it was reported, Facebook responded that it didn't violate its Community Standards. Why not? After more than a month of continual complaints, Facebook relented: it had only looked at the photo, not the text that went with it. The post was removed.

On 2 July, there was a disturbance in Mandalay, the second largest city.[20] Two Muslim teashop owners were accused of raping a female Buddhist employee. The claims – which were false – had first appeared on a blog, but then turned viral once repeated on Facebook.

Over the next two days, two people died and ten were injured as mayhem spread. Concerned that the violence could reach nearby towns that had also seen religious conflict the year before, the president's office tried to contact Facebook to get it to block the content. But there was no local office, and the company had no nominated contact in the region. Instead, Zaw Htay, a senior official for the president, contacted Chris Tun, then working for the Myanmar arm of the consultancy Deloitte and well connected in the local tech community, and pleaded with him to use his contacts to persuade the company to act. Tun tried in turn to get in touch with someone who could take action at the US head office, but the ten-hour time difference between Myanmar and California proved insurmountable. The president's office took the quick way out, and temporarily blocked the entire country's access to Facebook. A spokesman for the government later said that cutting off access worked: the unrest died down. (A curfew in the city and neighbouring towns may have helped.)[21]

Tun awoke the next day to find emails from Facebook staff awaiting him. But they weren't about the violence or the controversial content; they were concerned about Facebook being unreachable. The site being online outweighed any other consideration – not that they seemed to be aware what other considerations there might be, despite Callan's efforts the previous year.

In retrospect, Tun's problems weren't surprising. Though Facebook didn't admit it at the time, a single Burmese speaker was in charge of moderation for the entire country. But they weren't in Asia; they were in Dublin, Ireland, where for various jurisdictional reasons Facebook's non-US headquarters are located. That person mostly relied on Facebook users flagging content as unsuitable. Two million users might generate enough to keep one moderator busy, even if they only reported content by accident, but the moderator also faced the challenge of trying to search through Zawgyi posts using a computer outside Myanmar that only understood Unicode. In effect, Myanmar was a classroom of millions of excited, rowdy children without a teacher to keep discipline.

This was not 'social warming' as we might understand it from looking at other countries' reactions to the arrival of smartphones and social networks, where there would be subtle changes in behaviour as populations already familiar with social networks and computers adapted to an always-connected existence. This was a country emerging from years of censorship, with no preconceptions about computers, discovering what seemed like the whole country, and the whole world, in their hands. This wasn't warming so much as incineration.

Later in July, the UN's special rapporteur Yanghee Lee wrapped up a ten-day visit to the country by noting that polarisation between Buddhists and Muslims was growing.[22] 'I am concerned by the spread of hate speech and incitement to violence and hostility in the media and on the internet, which have fuelled and triggered further violence,' she said, and called for legislation that would target hate speech, though not speech generally.

Just over a fortnight after the debacle of Tun's fruitless attempts to connect with Facebook, a meeting was convened in Yangon bringing together the government, Google and Facebook's director of policy for the Asia-Pacific region, Mia Garlick.[23] She told the audience that Facebook would get the site's code of conduct and user guidelines translated into Burmese more quickly. In fact, it took another fourteen months.

Over the next two months the new mobile operators, Ooredoo and Telenor, opened for business, competing ferociously on price and offering cheap data plans to attract new users. As had happened in so many other countries, having a smartphone was quickly shifting from being a luxury to being a necessity. Ooredoo's edge was to offer free Facebook access to people who bought its SIMs.

Social media began to emerge as a key player in unrest. In the past, before mobile phones, an incident could occur but the reaction could be quickly quelled, and wouldn't spread far, nor linger. Now amplification and distortion of what had happened would continue to echo throughout Facebook, far beyond the reach of the lone moderator six time zones away. Threats sounded more immediate because they were on a phone in front of you. Insults felt more personal. Everything was more dramatic than in a world of state-controlled radio, censored newspapers and minimal connectivity. Facebook's mission to 'give people the power to build community and bring the world closer together' doesn't sound so beneficial when the community is carrying machetes, and they're being brought closer to people they liken to rats.

In 2015, twenty million people in Myanmar signed up to get mobile service for the first time – a bigger number than for any other country except China and India, which could boast populations twenty times larger.[24] The growth in ownership had far exceeded even the more optimistic of the Deloitte forecasts, which had only expected it to hit 50 percent by 2016.

The internet – more precisely, Facebook, since it was the internet to almost everyone – was also starting to figure as a significant media source. Polling by the International Foundation for Electoral Systems (IFES) in 2015 found 22 percent using the internet as one of their media sources, though it was still smaller than TV, radio and pamphlets.[25] Notably, one third of those under thirty-five used Facebook at least once a week. Later the same year, another IFES survey found 14 percent of the population saying they used Facebook every day, with a further 7 percent using it at least once a week. (For comparison, 41 percent said they watched TV every day.) Unsurprisingly, the heaviest users of Facebook were those aged between eighteen and twenty-nine.

Despite that rapid growth, even by early 2015 Facebook had just two Burmese speakers reviewing posts reported to them. They weren't Facebook staff, but employees of the company that had been hired to handle moderation.[26] Nor were they based in Myanmar.

People on the ground, meanwhile, were increasingly concerned at the content they were seeing in postings online. In March 2015, Schissler gave a talk at Facebook's headquarters in California about Facebook's role and responsibilities in Myanmar's problems. In May 2015, David Madden, a co-founder of a local tech incubator company called Phandeeyar, also travelled to California and told Facebook's senior executives that the risks emerging in Myanmar were as serious as Rwanda in 1994. Then, radio stations had instructed citizens to carry out the killings that amounted to genocide of the Tutsi; at least half a million died. Facebook risked being to Myanmar what radios had been to Rwanda, Madden told them: the platform whose motto was to move fast and break things was moving too quickly and breaking the uneasy social compact. Among those attending the meeting was Garlick. Possibly she was getting used to hearing the message Madden brought.

In a subsequent email to Madden, Facebook assured him that his warnings had been shared internally and were being taken seriously.

Later that year, the journalist Craig Mod visited Myanmar with the consultancy Studio D, seeking to find out how rural smallholders were using smartphones, hoping they might be a vector for education. He found the 'burner system' for setting up Facebook accounts was well established in rural areas: nobody he spoke to understood the idea of an email address, let alone used one. 'You realise most contemporary apps require an email address to sign up or create an account through, and it's a big ask for a completely computer-illiterate society to even understand what an email address is,' he told me. 'To understand the topography of writing the dots and the @ symbol – it's very weird when you distance yourself from it.'

Some of the farmers were near-addicts: they topped up their $20 pay-as-you-go phones with shop-bought scratch cards whose code unlocked 25-megabyte chunks of data. Some would use

500MB in a month – the same as a British smartphone user at the time.[27] Others used far more.

If someone lost their phone, or forgot their login, they'd just get another phone with its burner Facebook account, and start again from scratch. Some people would have multiple accounts that they used simultaneously; often people would share login passwords. Because – why not? The idea that a Facebook account had any value, in terms of personal content you wanted to keep to yourself, didn't exist. There's no exact word for 'privacy' in Burmese,[28] so the details about privacy settings would have been meaningless to those who didn't know English – not that they are much more comprehensible to those who do.

And people simply didn't know what was and wasn't true. All Facebook's assumptions about accounts, security models, personal profiling, threat models – all buckled and broke in the face of a population that came to it with none of the same presumptions or precursors. Being told that your Facebook account has been suspended for repeated content violations isn't much of a threat if there are five more logins immediately available on your phone.

The other basic assumption built into Facebook's model – you have friends! You want to stay in touch with your friends! You want to hear from your friends! – also collapsed. 'We noticed a lot of the farmers weren't [Facebook] friends with anyone in their village, and we'd ask them, "Why aren't you [Facebook] friends with your cousin?" And they'd go, "Why would I be friends with my cousin? I see him every day,"' Mod recalls. 'That concept of using a social network to stay in touch with people you knew, at least in the groups we spoke with, was alien. It seemed kind of useless.'

In fact, they simply didn't see it as social, and didn't perceive the idea of the 'network', Mod says. 'Having had no experience and no exposure to networks, to them all it is, is a delivery mechanism to get news. It just became this easy, quickly spreading thing that was a way for farmers to basically read the news, see pictures of five-legged cows, and because up until that point news equalled trust, I don't think anyone took a second to question the validity of the sources.'

Truth, lies, entertainment, incitement: there wasn't a distinction. The gatekeepers and the gates that had been in place for decades were swept away in a flood of pixels and notifications and screens with bright moving colours in people's hands. 'It was just like mainlining entertainment, essentially, for these people,' says Mod. 'They went from no TVs or anything like that, to instantly having this pipeline of pretty cheap data and essentially infinite entertainment.'

As a journalist, Alan Davis reported on the ethnic cleansing in the Balkans, and first visited Burma, as it was then known, in 1990. He returned repeatedly, and in 2012 had visited again in his role as the Asia director of the Institute for War and Peace Reporting (IWPR), a non-profit that tries to improve reporting in conflict zones, closed societies and countries shifting to democracy. Myanmar was an obvious candidate, and on that 2012 visit he, like Callan, had been concerned by what he saw, both in the violence and in the journalists' reporting, which repeated hate speech ('they couldn't understand what hate speech was, but that's another story'). The IWPR decided to get funding to train local journalists in how to recognise inflammatory speech, and to monitor what was going on. That took more than two years; he returned in 2015. A researcher on the ground interviewed citizens and found, as Davis recalls, that 'it's not the traditional media you need to be worried about. It's pretty much leaflets and CDs [of sermons], and also look at social media.'

Davis helped set up a group of local journalists as monitors of output on TV, radio, printed media and online. 'We found right from the get-go that all the hate speech, all the nasty stuff was on Facebook, so 95 percent of our time was monitoring Facebook.'

One of the first reports was a one-minute clip of a man who had filmed himself talking to his phone while driving, insisting he had learned of a Muslim Bangladeshi plot to blow up the Shwedagon Pagoda. 'Nobody has any idea who this person is, but it just went on Facebook, then it was shared around and around, and it steamrollered.' Such incidents happened repeatedly: 'It fed off other incidents,' Davis recalls. 'When there was the bombing down in Nice [France, in July 2016] you would see a spike [in hate speech about Muslims] subsequently on Facebook.'

The shift from closed to open, from authoritarian to democratic, and the existence of a dominant ethnic group all made trouble inevitable, Davis says. 'Democracy is a messy business, and this is democracy for the majority – and minorities in countries like Burma or Yugoslavia are always going to suffer because they don't have the voice. It's always going to be the powerful, after you change from totalitarian to open, who are going to have control. And it's going to get very violent.' And that's without social media, he points out: 'When you add those in, it's more petrol.'

The introduction of Facebook Basics, which meant people could browse Facebook as much as they liked without being charged for data by their mobile carrier (though they couldn't see video or photos), probably made things worse, he suggests: 'It meant you incubated everything, and necessarily the bad stuff.' Access to non-partisan news sources cost money: 'You couldn't get Reuters or AP [the Associated Press], you couldn't get CNN on those phones, so in retrospect that was a lousy thing to do.'

Meanwhile, phone ownership and Facebook use kept rising. So did the tension. By the start of 2016, 80 percent had mobile phones – fulfilling the government's 2011 ambition – and 40 percent had mobile broadband, with the network company Ericsson forecasting that both figures would pass 100 percent by 2021.[29] Mobile broadband penetration, at around 30 percent, was already higher than Bangladesh, and level pegging with Vietnam, even though

Vietnam's mobile ownership was over 100 percent (some people had two phones) and Bangladesh's 80 percent.

Facebook was also well established. Querying its database for Myanmar-based users suggested, improbably, that there were more users there than in any other country in south-east Asia. Yet a representative survey at the time found that fewer than one in five mobile owners knew how to set up a Facebook account, or to change the default settings on any app.[30] The survey also found 41 percent of users used multiple accounts.

Despite the assurances of forthcoming action given to Callan and Madden in the preceding years, an article in the *Myanmar Times* in August 2016 noted that the online temperature in the country seemed still to be rising. 'There's a dark side to Myanmar's recent internet revolution,' began the report by Nick Baker.[31] 'Behind the selfies and the stickers, social media platforms like Facebook have become fertile new ground for hate speech – with an audience growing by the day.' Facebook's local user base had now passed ten million monthly users (measured by Facebook as separate accounts that logged into it at least once a month), or just under 20 percent of the country.[32] But given the way that accounts were shared between people and on phones, it's hard to know if that number was an under- or over-count.

What human rights organisations observed was that the lack of digital literacy was being exploited to push hardline views. 'The number of shares, Likes and comments on false information or fake news is astounding,' Ma Zar Chi Oo of a literacy organisation told Baker. Old video was re-captioned as if it were brand new, claiming riots had broken out. Comments were inflammatory. Hate speech had a new, and comfortable, home. The American insistence that the answer to hate speech is more speech was about to be put to an empirical test.

* * *

Among those who benefited most from pushing extremist views was Wirathu. While more senior monks disapproved of his anti-Muslim messages, and preached reconciliation, he benefited from the amplification that Facebook's algorithm brings to those who get the strongest reactions.

Speaking to Buzzfeed News in late 2016, he said that his Ma Ba Tha movement – recognised by many as stridently anti-Muslim – owed a lot of its success to Facebook. 'If the internet had not come to [Myanmar], not many people would know my opinion and messages like [they do] now,' he told Sheera Frenkel, then Buzzfeed's reporter. 'The internet and Facebook are very useful and important to spread my messages.' Books and sermons were fine, but the internet was 'a faster way to spread the messages'. He recounted how his first account had been deleted by moderators almost at once; the second grew rapidly to 5,000 'friends'; and then he set up a network of Pages, which two full-time staff updated every hour. Facebook would shut them down, he complained, but he was able to keep ahead of them.

His methods worked. Frenkel met one tour operator who had previously heard vaguely about a hardline monk, but then seen Wirathu's content on his son's Facebook account, and become convinced. He brandished faked and miscaptioned pictures of Isis in Iraq, captioned to suggest they were killing Buddhists. He showed her a post about a Muslim plot to blow up Buddhist sites. (As the police later explained to her, this is a recurring urban legend in the country, used to foment anti-Muslim feelings. Facebook gave it fresh, incontestable life: how could you disbelieve the evidence that was on the *internet*?)

In March 2017, the government banned Wirathu from giving public sermons anywhere in Myanmar for a year, after he praised the killers of a high-profile Muslim lawyer. But the ban didn't proscribe social media, nor prevent other people reposting his content, or sharing things he had previously said. Though Wirathu sat in sermons with tape over his mouth as a protest, the reality was that he was still free to reach his biggest and most receptive audience.

In spring 2017, the IRI (whose motto is 'advancing democracy worldwide') surveyed people's news sources in Myanmar again.[33] Now, 25 percent used Facebook, of whom nearly three-quarters used it almost every day and liked it most for finding international or local news – far ahead of other uses such as 'chatting with children/friends'. And compared to the 2014 survey when just 3 percent used the internet for information, now almost 40 percent said they got 'most' of their news from Facebook, and another 35 percent got 'some' from it. Only 10 percent said they rarely or never did.

Significantly, a quarter of people said they saw posts about ethnic and religious conflict every day; another third saw such posts at least once a week. But they wouldn't tell people not to share such content; two-thirds said they never did, and another one in five only 'sometimes'.

And was what they saw on Facebook true? A total of 60 percent thought that most or all of it was true; one-third thought most or all was false. (The remainder didn't respond.)

A clear picture emerges: the population was switching from getting its news from TV, papers and personal conversation to a disintermediated, unregulated free-for-all where it's impossible to know what's true, but where ethnic tensions are running high, and getting higher.

In early 2017, Madden tried again to drive home the point to Facebook that he thought he had made in May 2015. He and a number of local NGOs met once more with the company and argued – yet again – that the content moderation system simply wasn't working. The fear was that as the malice simmered, it would boil over into something much worse. Madden later recalled in a PBS Frontline documentary that Facebook's team said they would need to 'dig into this' and 'come back with something substantive'.[34] Madden says: 'The thing was, it never came.'

In June that year, the number of anti-Rohingya messages being posted by Ma Ba Tha supporters on their Facebook began to ramp up. At the beginning of August, the intensity increased noticeably.

Facebook's system wasn't equal to it, though. Reuters tested its Burmese-to-English translation tool with a Burmese phrase posted on the site in August 2017. The English translation it offered was: 'I shouldn't have a rainbow in Myanmar.'

The original Burmese phrase, confirmed by a native speaker: 'Kill all the kalars that you see in Myanmar; none of them should be left alive.'

On 25 August 2017, militants from the separatist pro-Rohingya ARSA attacked a number of outposts in the north of Rakhine state. In the town of Maungdaw, twelve security personnel were killed. The attackers targeted Hindus. In response, Myanmar's military began a fateful series of attacks against the Rohingya, in what was called a 'clearance operation'. Hundreds died in vicious killings that left families divided or slaughtered, and entire villages were razed or bulldozed. Tens of thousands fled west to Bangladesh. The army was attempting to commit genocide out of view of the modern world. But satellite pictures, smartphones and communications networks meant that this time, the international media could see and quiz the perpetrators.

The army had its own response to media interrogation, however: disinformation. Within hours of the military reprisals, more than 4,000 Twitter accounts began tweeting pictures and videos from the conflict, blaming the Rohingya and calling them '#bengali from #bangladesh #terrorist'. They posted faked photos suggesting the Rohingya were burning their own villages – a tactic to cover the military's own role in the conflict and to paint the victims as liars. Meanwhile, the number of anti-Rohingya posts on Facebook by Ma Ba Tha supporters spiked again.

The Twitter accounts were what first intrigued Serrato.[35] 'It was odd, because you never saw any kind of Burmese community on Twitter,' Serrato recalls. When he investigated, he found they had all been created just two months earlier, in June. 'So that was the first trigger – you had thousands of accounts which had popped up and were immediately accusing the Rohingya at large for the deaths of these security personnel.' He realised that the target of the tweets wasn't the Burmese population: 'Most of the messages were in English, so they were targeting international audiences, they were adding journalists in their mentions [so they'd come to journalists' attention]. It seemed to be a propaganda campaign which was partly driven by trolls, and automated accounts as well.'

Inside the country, Facebook was being used to incite anger to justify the army's – and by extension the government's – actions. Serrato and his team collected data from Facebook Pages of Groups that Wirathu belonged to and analysed it. A handful of the members – less than 1 percent – generated nearly a third of all interactions in the Group. 'The Ma Ba Tha group [posting] activity specifically was driven by maybe ten to twenty members, and it appeared in a coordinated fashion. They had certain hours where there was a lull in activity, which suggested to me that it was an active operation.' Serrato, with six years' military experience, recognised a pattern in the posting hours: 'like they were working, and then they were taking breaks and having a shift change, and then began working again.'

The content was intended to outrage and radicalise. Some posts suggested mosques were being used as arms dumps to blow up Buddhist pagodas.[36] Not everything was extreme; it didn't have to be. 'There was sometimes content which wasn't necessarily outright hateful, but it was still intended to incite negative sentiments, particularly about the Rohingya,' Serrato says. Sometimes there would be a specific call to violence, but that was rare: 'What you usually saw is a lot of moderate to fringe material which was intended to keep boiling the pot. Just keep it simmering.'

Serrato is sure of one thing: the murderous army attacks would have happened anyway. What was different was how the army sought to portray and justify its actions, and how the population, now primed to believe that the Rohingya were part of a vast plot, reacted. Unlike 2007 or before, it was now in a goldfish bowl, with international scrutiny. 'I think what's clear is maybe they wouldn't have had the public support they did, absent the kind of manipulation that they used the platform for,' Serrato explains. 'I mean, they reached millions more people.'

If we're trying to evaluate a concept like social warming, how do we measure the social temperature of a country? One measure is self-evident: the frequency and seriousness of internal conflicts, whether military or just interpersonal. In Myanmar, the events that followed the gradual opening in 2010 all pointed to a rise in that temperature. The way that outrage was amplified, and spread around, was frequently aided by the combination of social networks – specifically, Facebook, and especially its combination of a laissez-faire attitude and puzzlement over the *sui generis* nature of the challenge of moderating Burmese content – and smartphones being eagerly adopted by a population that was new to untrammelled content.

One problem in trying to measure with any precision how that temperature changed at the personal level is the paucity of polling data. Few organisations existed in Myanmar to ask questions of the population. Most polling only sought people's views of upcoming elections, and whether they would trust the result.

There were a few indicators. A telling survey in November 2017 by PACE (People's Alliance for Credible Elections) Myanmar found that there was more discomfort about the idea of having a Muslim as a neighbour (69 percent) than someone with different political views (39 percent). The distrust of Muslims was notably higher in rural areas. And generally, only one in twenty said they'd be happy if their sibling married a Muslim; more than two-thirds said they'd be uncomfortable with it. There was also the Fortify study in 2016 that found a correlation between anti-Muslim content and outbreaks of violence.

One key difference between the period before 2010 and the succeeding years was that Facebook increasingly brought such anti-Muslim content in from the outside world. The 'Londonistan' content that Serrato noticed would essentially have had no avenue into the country a few years earlier. But via Facebook, as more people responded to it, the algorithm would show it to more people as 'news you might be interested in'.

Sheera Frenkel, for Buzzfeed, noted that it was often those who posted the most, and had the wildest stories – whether true or not – who would be at the top of the News Feed.[37] Or, as a nineteen-year-old student who had just got her first phone explained it to her, 'It takes people who say crazy things and makes them famous.'

Myanmar, meanwhile, was making Facebook infamous. On 12 March 2018, the chair of the UN's fact-finding mission told the media that social media (which in Myanmar simply meant Facebook) had played a 'determining role' in the killings of the previous August: 'It has . . . substantively contributed to the level of acrimony and dissension and conflict, if you will, within the public. Hate speech is certainly of course a part of that.'[38]

Yanghee Lee, the UN's special rapporteur, commented that ultra-nationalist Buddhists had used it to incite violence and hatred: 'I'm afraid that Facebook has now turned into a beast, and not what it originally intended,' she said.[39] Facebook responded that there was 'no place for hate speech' on its platform and told the BBC that it took the matter 'incredibly seriously' and had worked with experts in Myanmar for 'several years' on 'safety resources and counter-speech campaigns'.[40]

In the first week of June 2018, Facebook once more sought to build bridges with the local tech community, holding a three-day meeting in a local hotel to discuss the challenges. 'We said, "Come and actually learn and listen to what happened,"' IWPR's Alan Davis recalls. 'They sent a whole bunch of people, not just from Singapore [Facebook's regional headquarters] but from Palo Alto and Washington.'

And what was the effect? 'They just didn't get it at all,' Davis says. 'Rather than actually listen to the issues, the challenges, to what we had found, the big things, they were kind of "Oh, that's interesting, now let's go into groups and we'll facilitate this and facilitate that and, oh, OK, safety and security, let's talk about that and how to stay safe online."' He's almost speechless at the memory.

To cap it all, the person leading the Facebook delegation was Sarah Oh – who from 2014 to 2016 had worked for Phandeeyar, one of the organisations that for years had been imploring the social network to take the problem seriously.

'What frustrates me is that a lot of these things were foreseeable, even without social media and Facebook,' says Davis, reflecting on the amplification of social tensions in a country that hurtled from decades of authoritarianism to democracy in just a few years. 'It makes me laugh, the whole thing of Zuckerberg saying, "We bring people closer together." Do they heck. Yes, closer together to stab each other.'

* * *

What about Facebook's insistence that it treated the subject 'incredibly seriously' and that there was 'no place for hate speech' on the platform? That wasn't borne out by an investigation by Reuters in August 2018, which found more than a thousand examples of comments, some up to six years old, such as one saying the Rohingya were 'non-human kalar dogs, the Bengalis ... we need to destroy their race.'[41] Violent, dehumanising, pornographic: as Reuters pointed out, the content clearly broke Facebook's rules. So how was an enormously profitable company still allowing it, after being warned for years informally and formally?

In the same month, the UN published a report into the genocide from a year before, and noted that Facebook had been 'a useful instrument for those seeking to spread hate'. Buzzfeed News analysed 4,000 posts made on Facebook between March 2017 and February 2018 by politicians in Rakhine state, and found that 10 percent of them contained hate speech.[42] Facebook acknowledged being too slow to remove content, but said there were now sixty people moderating Burmese content and the aim was to grow that to a hundred. (Buzzfeed found that the recruitment seemed to be happening in Dublin – hardly a well-known location for Burmese diaspora.) Facebook also said the ongoing problem was too few people reporting content – demonstrating once more two flawed yet embedded assumptions: that people who had lived under a military dictatorship would report fellow citizens, and that they would do so for content they probably agree with, even if it's hate speech.

Facebook chose the evening of the US midterm elections on 6 November 2018, when the news focus was on a huge reshuffle of American politics rather than technology companies' actions in foreign lands, to make a number of announcements about its work in Myanmar, and release an independent report by a global non-profit consultancy called BSR into its role there.[43] It said support for posting on Facebook using Zawgyi would be dropped, and everything would be shifted to Unicode; for those using Zawgyi, programs called font converters would mean that postings on Facebook itself would always be Unicode-based, but those using Zawgyi-based equipment would see it translated back.

Facebook also said it didn't plan to open a local office in Myanmar, because the government could use the leverage of threatening to seize computer equipment or 'place Facebook staff at safety risk' to influence local policies. As observers quickly pointed out, that hadn't been a problem in Thailand, or Indonesia, or the Philippines, in each of which there were similar pressures. The job of spotting content and posters with malicious effects continued to be, in effect, outsourced to journalists and NGOs.

BSR's report confirmed what had been said for years by multiple people, privately and publicly: Facebook was being used to 'spread hate speech, incite violence and coordinate harm.'[44] Part of the problem was that 'the Facebook platform and Community Standards rely on certain legal, political and cultural assumptions (such as freedom of speech and rule of law) that do not hold true in the Myanmar context today.' Furthermore, 'a minority of users is seeking to use Facebook as a platform to undermine democracy and incite offline violence.' The guardrails of digital literacy that in other countries would prevent the site being used as a weapon simply didn't exist for users in Myanmar. Threats of suspension meant nothing.

'Internet companies will often be linked to human rights impacts that they do not cause or contribute to,' said the BSR report. 'For example, internet companies may be linked to hate speech, child sexual abuse material, and hacking that takes place over their platforms, even though they do not cause or contribute ... themselves.' This neatly swerved around the fact that the platform provides, well, the platform: some enable such content and actions more easily than others. If the car you're driving is spewing out more pollutants and ruining the atmosphere more rapidly than others because of what you've put into it, finding a fix isn't the driver's responsibility. It's the manufacturer's.

'Maybe Myanmar isn't ready for Facebook yet,' one interviewee commented to BSR.

In 2010, Myanmar's GDP was $49.5 billion.[45] By the end of 2018, it was 54 percent bigger, at $76 billion, a 5.4 percent compound growth; Myanmar seemed to be hauling itself out of the economic swamp. Perhaps Deloitte's forecast about the benefits of mobile had been correct.

But at what cost? Ethnic tensions had been strained up to and beyond breaking point, and the country had descended into genocide and ethnic cleansing. Hundreds of thousands of Rohingyas had been displaced, perhaps forever, into Bangladesh.

Nor do social effects always have to be that obvious to be negative. Anders Larsson, the chief technology officer for Ericsson in Myanmar, observed that when he had first worked there in 2013, preparing to build its mobile mast network, Yangon was 'pristine ... a lush, green city where the people were calm and relaxed'.[46] A few years later, he returned, and found that people were connected, better educated, had better jobs and greater freedom. But it wasn't the same: 'With the economic boost, pollution has followed,' he commented. 'The roads are jammed with cars that fill the air with thick, black fumes. Since the electricity grid is either poor or non-existent, the sale of large, industrial-grade diesel generators is booming. Jogging in the capital now leaves a layer of black dust on my skin.'

Pollution comes in many forms, some more subtle than others. Buzzfeed's Frenkel went back to the nineteen-year-old woman who had just got her first smartphone and had spent a week scrolling the ocean of the internet (or more accurately, Facebook), receiving Friend invitations from strange men in Thailand, taking photos to use as backgrounds, trying to figure out fake from factual. Frenkel asked her: would she recommend the internet to other people?

'Yes,' the teenager replied. 'As long as you know it might make you angry.'

In early September 2018, the No. 2 at Facebook, chief operating officer Sheryl Sandberg, testifying to the US Congress, said that 'Hate is against our policies and we take strong measures to take it down.'

But Facebook was warned repeatedly, yet didn't act. Either the organisation had become dysfunctional, incapable of reacting to warnings about the effects of its product, or was indifferent to them, putting a vague principle of 'connecting people' above the well-being of millions of humans at risk of persecution. Either way, the inevitable outcome of its presence in Myanmar was to accelerate a violent process. Hate might have been against Facebook's policies. But it inevitably became one of its consequences.

Aela Callan is still frustrated at all the warnings that were missed and signs that were overlooked. Facebook's reaction, she says, was 'But how could our platform, which is so wonderful, be doing something so awful?' Now she feels that an opportunity perhaps even to change the course of history was missed. 'I stood on a stage in Stanford in 2014 and said, "Myanmar is the Petri dish. This could happen anywhere."'

And of course it did happen somewhere else. It happens everywhere, to a greater or lesser extent, but it always happens where social networks touch down and gain significant traction. The US

– highly developed, educationally and technologically advanced, globally connected – seemed so unlike Myanmar – underdeveloped, digitally untrained, politically divided and without a free press – but Callan saw the inflammatory, socially disruptive effect of social networks take hold there too. Of her 2014 prediction, Callan says, 'I never thought in a million years it would happen in three years' time in the US. But it did. And they didn't listen.'

6

DIVIDED VOTE: HOW SOCIAL MEDIA POLARISES POLITICS

It was 3 September 2019, and Ed Vaizey was annoyed. This was unusual for the normally cheery MP for the constituency of Wantage. But after fourteen years he had effectively been kicked out of the Conservative parliamentary party, along with twenty others, because he had refused to vote for a bill backing a 'no-deal' Brexit (a negotiating threat against the European Union akin to breaking your own leg ahead of a race). The reason for his annoyance, though, was the discussion inside the Conservative MPs' private WhatsApp group.

'Remove these people from the WhatsApp group', the pro-Brexit MP Zac Goldsmith, representing the Richmond constituency, had written bluntly. Vaizey resented that deeply, and thought Goldsmith might have shown at least a little mercy.

'I'd been to campaign for him in the [2017] general election and I'd campaigned for him in his [2016] by-election,' he told me a little later. 'I thought that was typical behaviour of someone who doesn't want to heal a divide. But you can also see it as a kind of off-the-cuff reaction, not thinking through who's watching and who's who on this forum.'

Being kicked out of the party? Not such a big deal. Being targeted in the electronic forum that every colleague would instantly see? A much bigger deal. Months later, the incident clearly still rankled. The political had become personal; a social media group visible only to Vaizey's workmates had enabled tensions to surface, and loyalties to be dumped. Before social media, such metaphorical knifings almost always happened in private. Here they were on show for everyone to see.

During his fourteen years in parliament, Vaizey saw first-hand how the practice of politics, inside and outside Westminster, became more and more inflamed by the arrival of social networking apps. His experiences were a microcosm of the broader changes that have happened as politicians, who are people after all, have had to adjust to the background roar of a citizenry able to make their feelings and opinions known not just at the ballot box, or in opinion polls, or in letters, but on social media in real time.

You can think of politics as a two-by-two grid: one square of the grid consists of politicians talking to each other; a second, politicians talking to the public; a third, the public talking to politicians; and the fourth, the public discussing politics. Social media has taken over all four, though served by different platforms.

For discussion between politicians, WhatsApp's inherent security, ease of creating and managing groups, and ability to work on any smartphone has made it a favourite. (In the US, the Signal app, which has almost all the same features but isn't owned by Facebook, seems to fulfil that function.)

In the other three quadrants, Twitter, Facebook and YouTube have become the principal method of discussion – although in some countries, notably Brazil and some African countries, WhatsApp's zero-rating for data has made it the prime conduit for large amounts of political discussion involving public and politicians. In the US, by contrast, WhatsApp is little used even in the general population: by mid-2018 there were only about twenty million users there.

* * *

WhatsApp's entry into British politicians' lives seems to have happened on both sides of the political divide in 2015, the year after Facebook purchased it for $19 billion. By then around a quarter of the UK population had installed it.[1]

Political parties in the UK have in the past been eager adopters of personal communications technology for their organisers. The Labour Party's success in the 1997 general election, and its communication strategy afterwards, relied on issuing pagers to every candidate, and sending messages from headquarters with the chosen talking points of the day. Pagers were one-way, though. The advent of smartphones and then private networking apps, particularly WhatsApp, ushered in a new way of communicating with each other. Given that politicians thrive on discussions in small groups, it could almost have been invented for them.

The app first gained popularity with the Conservatives in the May 2015 election, when advisers used it to keep track of journalists and demonstrators they wanted to avoid. After that election, plans to set up a party-wide group for the 330 Conservative MPs were stymied by the 256-person limit on groups, so new ones were spawned for the overflow.

As a longstanding MP, Vaizey had always been in the main WhatsApp group. Being removed was the ultimate disbarment, like a cashiered officer having his sword broken. As he remarked in a newspaper article after the December 2019 general election, at which he didn't stand, 'This being the twenty-first century, you haven't actually left until you have been booted out of the Tory MPs' WhatsApp group.'[2]

On the Labour benches, the smaller number of MPs – 232 – after the 2015 election meant everyone could be included in a WhatsApp group. Stella Creasy, then aged thirty-eight, who had been the MP for Walthamstow in London since 2010, pushed for its adoption. 'When I first got elected, I was so obsessed with how all these people in [parliament] are incredibly dysfunctional, because they're not speaking to each other, not interacting with each other,' she told me. The many conflicting calls on MPs' time and presence often meant choosing between working with their constituents, or meeting colleagues, or attending votes in the House of Commons. 'They don't know what each other thinks until a crisis moment comes and suddenly they're asked to vote a way that they don't agree with,' Creasy explained. Having a smartphone and a communications app in their pocket could change all that.

She thought that 'if they were able to connect with each other, they would somehow be kinder to each other.' In retrospect, the effects weren't all positive: 'One of my greatest regrets is teaching a lot of Labour MPs to use WhatsApp in the mistaken apprehension that one of the problems in our politics was basically people didn't interact with each other except at points of crisis,' she told me. 'Actually it's been the reverse. During the period of Brexit, WhatsApp groups have become especially unpleasant.' She laughed. 'So I bear some responsibility for that, because it was me who taught them all to do it, thinking it would be a good idea to get people talking.'

Many MPs were already keen users of Twitter and Facebook, as Creasy herself had been since before becoming an MP, when she worked for Involve, a think tank advocating greater public participation in order to strengthen democracy. Neither platform offered the immediacy, privacy and convenience of WhatsApp, though.

Creasy hoped then that WhatsApp might unite a party that was becoming increasingly divided after two election failures. 'I remember talking to various MPs, and saying very bluntly that with the way the Labour Party was at the time, we all needed to start talking, and here's this very simple way you can do it from your phone. And there were some conversations with people to explain who couldn't see what.' Privacy, after all, is essential to plotting; and what would politics be without plotting?

The prime minister in 2015 was David Cameron, the Conservative leader, who had been surprised by his victory. He had expected another five-year coalition like the one he had just been through with the centrist Liberal Democrats after failing to win a majority in the 2010 election. Instead, he found himself victorious, and committed by a manifesto pledge to a referendum on the UK's membership of the European Union.

Despite his relative youth – he was forty-eight – and his predilection for 'chillaxing' by playing trivial games on an iPad, Cameron was not initially an enthusiastic adopter of social media. 'The trouble with Twitter, the instantness of it – too many twits might make a twat,' he quipped on a breakfast radio show in July 2009.[3] Under pressure from his advisers, he relented and posted videos to YouTube made in his kitchen, before eventually joining Twitter in October 2012. Predictably, his first tweet ('I'm starting [the party] Conference with this new Twitter feed about my role as Conservative Leader. I promise there won't be "too many tweets" . . .') was met with a slew of abusive replies.[4] Nor did his inner circle adopt WhatsApp in 2015, even as others around the party were beginning to realise how they could benefit from it.

As the momentum began to build towards the referendum in June 2016, Cameron, for all his modernism, used traditional methods of political persuasion: barbecues, lunch parties, the glamour of visits to Chequers (the prime minister's official country residence). 'No one in Cameron's team ever thought, "Let's get every single pro-Remain MP in the Tory Party on a WhatsApp group and feed them lines constantly so they can go out to bat for us,"' Vaizey says a little wistfully, perhaps imagining what might have been. 'And nor was there a Conservative MP group where David

Cameron could occasionally appear and say, "Listen, guys, it's all very painful, I know, but let's try and stick together."'

Meanwhile, pro-Brexit MPs began organising in their own WhatsApp group, for the so-called European Research Group (ERG) – an innocuous name for a group of about seventy Conservative MPs whose aim was to acquire a majority vote for Brexit.

After that majority had been achieved – by methods that, as we'll see later, relied heavily on social media to sway voters – Cameron departed, to be replaced by Theresa May in July 2016. The influence of the ERG, and particularly its WhatsApp group, grew greater and greater. Within months, the group was influencing the government's decision-making on Brexit policy. By February 2017, discussions in the WhatsApp group had pushed members to publicly criticise high court judgments that didn't favour their position, and attack forecasts by the governor of the Bank of England that they felt were insufficiently optimistic about the impact of Brexit.[5] They coordinated attacks on the chancellor and prominent centrist think tanks that they also felt were too pessimistic about Brexit. The group was so large and influential within parliament that the prime minister's office felt obliged to negotiate with, and even give advance policy briefings to, them. Otherwise it could use its leverage to sway public discussion – often led by the Twitter feed of Steve Baker, then the ERG's leader.

By 2017, politicians of all hues had moved onto WhatsApp as the preferred venue for their private discussions. Everyone had seen the fate of Hillary Clinton in the US presidential election the year before, when leaked emails – those from her work, those from the Democratic National Committee, and finally those from her campaign chairman John Podesta – had become the focus of story after story. Email, MPs realised, was simply too insecure as a messaging option: it was too easy to forward, or misaddress, or to be hacked. WhatsApp, on the other hand, could not be hacked.

Even while MPs would frequently criticise the encryption of the app in public if it was used by a terrorist organisation – 'It is completely unacceptable. There should be no place to hide,' said the Conservative home secretary Amber Rudd in March 2017, after what seemed to be a 'lone wolf' terror attack in London – they were very happy to reap its benefits for their own schemes.

The Tory WhatsApp group was only the beginning, spawning many smaller ones organised around the ideologies that were becoming more intensely polarised within the party.

By the time May called a general election in April 2017, WhatsApp was entrenched in British politics, with MPs having multiple separate groups – extreme Brexiters (the ERG), Labour MPs, female Labour MPs, moderate Tories, and so on. The talking points to use in media interviews, known as 'lines to take', could be much more easily transmitted, and feedback – so those lines could evolve – also given.

Inside the Labour Party, WhatsApp had also become essential. In 2017, centrist activists inside the headquarters set up a WhatsApp group to plot against the then-leader Jeremy Corbyn ahead of the election, trying to find a way to dislodge him.[6] He remained deeply unpopular with many of his own MPs, and subgroups multiplied inside the parliamentary party, all fed and watered by WhatsApp messaging.

'It's that group thing, isn't it,' Creasy remarked to me, drawing on her own academic study of social psychology. 'The Bay of Pigs effect, where we all get together, reinforce each other's view that what we're doing is the right way to deal with it.' Often that would reinforce discontent among those opposed to Corbyn's leadership: 'There would be a bunch of people who genuinely agreed that, say, the way Jeremy Corbyn had approached the situation in Salisbury was completely wrong,' Creasy says. (Corbyn refused in March 2018 to condemn Russia for the attempted assassination, using the nerve agent Novichok, of two people living in Salisbury.) 'Were you to put in a message about being pro-immigration, there would be huge divisions. But they would all agree that whatever the answer was, it wasn't Jeremy Corbyn.'

The 2017 election was calamitous for May: instead of boosting her authority and majority (which might also have diminished the ERG's leverage) she lost both. To pass legislation, she was left reliant on the cooperation of one of the smallest parties, from Northern Ireland, and on every one of her own MPs. That made the ERG even more powerful. And now British democracy was about to fall victim to a small-scale, yet very effective, version of social warming: the tendency of a closed group of like-minded people to edge, gradually but relentlessly, towards the most extreme views held within the group.

The idea that social warming might affect how politicians go about their dealings with each other may seem far-fetched. However, it was completely predictable once they had a self-selecting group that had formed around shared belief in a topic. Experimental work pre-dating all the social networks now in use explains how.

In 1999, Cass Sunstein, then a law professor at the University of Chicago, carried out a number of social experiments that demonstrated what he called 'the law of group polarisation'. His seminal paper on the topic begins: 'In a striking empirical regularity, deliberation tends to move groups, and the individuals who compose them, toward a more extreme point in the direction indicated by their own predeliberation judgments.'[7] That is: groupthink is real, and drives the group towards the extreme of its common thinking.

In the experiments, a group of people who already favoured gun control would, after discussion with each other, support the idea with even more enthusiasm; ditto groups in favour of paying the minimum wage, or in favour of action on global warming. 'This general phenomenon has many implications for economic, political and legal institutions,' Sunstein observed. Why groups act that way is clear enough: we tend to be reluctant to go against the flow in a group, and if you all agree about something already, then shifting the whole group's view towards a more intense form of agreement is easier than shifting it away.

Thus if you stick a bunch of terrorists into a connected group, they'll agree that their cause is not only just, but essential, and therefore requires action; they'll come up with newer, wilder ideas for terrorism. If you put a bunch of Brexit-eager British MPs into a closed, connected group, they'll back more and more extreme views about how to enact Brexit. Or, as Creasy acknowledged, anti-Corbyn MPs will become ever more vociferously anti-Corbyn.

That played out as ERG members stiffened each other's resolve through a series of government votes on Brexit, in which they defied the party leadership again and again, sinking the deal that had been negotiated.

Occasionally, messages from inside the group would leak out to well-placed journalists. Some revealed how outrage (inevitably) would spread among the members. At the end of 2017, one posted a message to the group complaining about the 'disgraceful inaccuracy' of a factual error during a Sunday-morning news discussion programme on the BBC.[8] A cascade of angry support followed. By the eighth message, the hardline Brexiter Iain Duncan Smith was insisting that 'everyone on this [group] should lodge a complaint with the BBC. The editors of the programme will have known what was said and they should have corrected it.'

So what was the appalling error? In an off-the-cuff remark, the radio presenter had wrongly 'corrected' one of the guests by saying that the Labour peer Andrew Adonis had previously been an MP. As Adonis had previously worked for five years as a government minister, a job usually carried out by an MP, the presenter's momentary lapse was understandable, and probably not noticed by more than a few listeners. But to the Brexiters, whether or not they had heard it, the mistake was a sign of 'disgraceful inaccuracy', over which the programme editors should have immediately intervened.

While this seems like a classic case of Sayre's Law – that the outrage is greatest when the stakes are lowest – it's also an example of how quickly an extreme viewpoint can bubble up in such groups, and how reluctant participants are to shoot it down, lest they appear a less resolute supporter of the greater cause. If a prominent member is insisting everyone lodge a complaint, who's going to suggest that it's not that important and that everyone should enjoy their Sunday?

The ERG continued its attack on May, leading to an attempt to unseat her as leader in 2018, and finally succeeded – after thwarting her Brexit legislation again and again – in helping to install occasional member and apparently enthusiastic Brexiter Boris Johnson as party leader. WhatsApp had done its work on British MPs.

What about the public dialogue between politicians and their constituents? The first MP on Twitter appears to have been Lynne Featherstone, a Liberal Democrat MP who joined the service in March 2008, four and a half years after she became the first MP to take up blogging. A month later, she was the first to tweet from inside the Houses of Parliament, saying she was waiting to speak on a law debate. Observing this, Mark Pack (then a Liberal Democrat activist, and subsequently the party's president) commented, 'Imagine, perhaps, a future that includes MPs blogging or posting Twitter updates as, say, the Prime Minister answers questions at PMQs [Prime Minister's Questions]. It would add a whole new perspective to the event.'[9] Eight years later, stopping them from posting Twitter updates or livestreaming the event was all but impossible. MPs were addicted to the potential that Twitter offered to get their message out to their constituency – even if that might not be the same as the constituents who they were meant to be representing. While Featherstone's Twitter following stuck at around 20,000 (there are about 76,000 people in her constituency), MPs such as Labour's Diane Abbott, the Green Party's lone MP Caroline Lucas, and the former Labour leader Jeremy Corbyn had hundreds of thousands, or even millions, of followers.

Even so, your MP was now always there, always available (in theory) to be quizzed about anything and everything. Politics was transformed from being an occasional spot on the evening news or morning newspaper into a permanent availability where, once again, the battle for attention mattered. Responding to the issue *du jour* was quicker through a tweet or Facebook post than issuing a press release and hoping a journalist might see it. Twitter in particular became a parallel world to the rolling news channels that were hothousing politics.

Social media became the new soapbox, but with a twist. Twitter amplifies the personality of the person tweeting, and even though many MPs (and certainly ministers) would use younger, more social media-adept recruits to tweet on their behalf, some simply used it as a conduit for whatever was happening in their head. The results weren't always good, because there's no guarantee that politicians will be smarter or know more than the general public. On specific topics, it's almost certain that somebody out there will know more about a particular topic than any given MP. In theory, that should be a huge benefit of Twitter or Facebook. In practice, though, sorting the wheat from the chaff – some of it vitriolic – could be overwhelming.

MPs who had thought that Twitter would be the ideal medium to broadcast announcements of good deeds done and factories opened in their constituency quickly discovered things didn't work that way at all. 'It's a toxic forum,' sighed Vaizey. But when he was a junior minister in the 2010 coalition government, he was quietly proud of having a larger Twitter following than any cabinet minister, and enjoyed the buzz of responses that came when he made a media appearance: 'If you appear on television, you're like the bear shitting in the woods – you haven't really appeared unless people are reacting on Twitter.'

He never used Facebook – 'I find the feed I get is full of people's rantings. At least with Twitter, they're limited. With Facebook, you get people posting just enormous amounts of shit, and I just could not be arsed to look at it.' However, his social media team took a certain delight in telling him about Facebook Groups, he recalled, 'such as "Get Vaizey Out" or "Ed Vaizey's Useless And Shit", which I was blissfully unaware of. Very good for my mental health.' Female MPs complained repeatedly that Twitter in particular was too slow to act over abusive tweets and messages sent to them, revealing a twofold problem: people felt free – even encouraged – to be rude, and the platforms did almost nothing about it.

Had politicians suddenly become more hated? Had the public's views become more extreme? Although public opinion of MPs had fallen since a gigantic expenses scandal in 2010, the biggest change was that interaction between the two could occur more frequently, more rapidly: a Brownian motion of continual collision between representative and voters (or non-voters) that kept heating up the debate.

Vaizey perceives 'an element of inbuilt confrontation' in Twitter: 'You want to take people on and start a row.' Creasy agrees, but adds an important caveat. In 2011, the year after she entered parliament, she was able to use Twitter to organise a political campaign to cap payday lenders' interest rates. 'I was able to bypass "traditional" rules of engagement with my colleagues because I was able to speak to a broader audience,' she told me. 'It wasn't just about being on Twitter. It was the fact we were following people back, having conversations, asking people to help.' That has continued into the present day: in 2020 she drummed up support online to push for a parliamentary amendment making misogyny a hate crime. On social media, volume can make MPs pay attention.

Other patterns have also continued, though. In 2012, after the government capped the rates that payday lenders could charge, a Twitter account began attacking her as 'mentally unstable' and a 'self-serving egomaniac'. The source was soon tracked to an employee at one of the lenders, who was acting without sanction.[10]

In 2013, Creasy was the target of rape and death threats, for which four people – none known to each other – were subsequently convicted. Yet even some of her parliamentary colleagues thought she had sent the messages to herself for publicity. The court cases persuaded them otherwise, but Twitter didn't improve. 'Within eighteen months, that behaviour just seemed to become a fairly standard thing. There's two things: the violence of the conversations. And then there's the sheer volume. What makes me sad now is when I think back to 2010. I followed people on Twitter who talked about Walthamstow, which is a bit like sitting on the back of the bus and hearing people talking about it and being able to tap them on the shoulder and say, "Hello, actually I'm your MP, can I help?" Those lovely elements of having a wider net that you can cast to have a conversation are lost to me now because they're lost among both the violence and aggression with which people approach anything, and the sheer volume of correspondence which means you can't keep up with it.'

Seven years later, she was still the target of abuse there. 'Twitter definitely has become really, really unpleasant,' she told me. 'Facebook has waves of it – Facebook seemed a bit defunct six or seven years ago. It's come back to being very useful because that's where a lot of local residents are. For a lot of the work I do as a local MP, I go to Facebook.'

You don't have to look far in politics to find a social-warming feedback loop in action. Both the in-group systems that politicians use to communicate, and the out-group systems by which they are publicly rewarded through social media, are clearly pushing politicians towards more extreme political and online behaviour.

There's a reward system in place: politicians who take extreme positions politically have more followers, and those who shift towards extreme positions gain more followers. The ones in the middle who seek compromise tend to be shot by both sides, and don't win the popularity (or vituperation) that their more extreme colleagues do.

In August 2017, an analysis by the US-based Pew Research Center found that highly ideological members of Congress had more Facebook followers than moderates – the definition of 'ideological' and 'moderate' being determined by their voting record, using a longstanding measure.[11] In the larger House of Representatives, the most extreme had about 50 percent more followers; in the 100-strong Senate, the extremes had on average almost twice as many as the moderates.

Why, though? Pew's researchers looked at what sort of posts the more extreme members were likely to share, and found they tended to express 'indignation or disagreement on political matters'.[12] Those posts then got more 'engagement' as measured by Likes, comments and shares. As we might predict, the more outraged a politician's post – crossing from simple disagreement to indignation, from 'this is bad' to 'something must be done' – the greater the response: twice as many Likes and shares, three times as many comments. Whether that's an artefact of Facebook's algorithm or a bigger response from the same audience doesn't matter;

the message for politicians considering what to post next time is clear.

For the politicians' purposes, it didn't matter whether people were agreeing or disagreeing in their comments or their sharing; in the attention economy, every moment when you can crowd out your opponents is a benefit. They hew faithfully to Oscar Wilde's observation that the only thing worse than being talked about is not being talked about.

Bill Brady, the Yale University researcher who examines how 'moral outrage' spreads on social networks, told me that 'There's no doubt that politicians, whether implicitly or explicitly, are aware of this general idea that if you post moral and emotional content, it can help diffuse through a network.' (He co-authored a paper in 2019 looking at tweets from Donald Trump, Hillary Clinton and the members of both houses of the US Congress: it found that 'moral-emotional' messages gained more retweets and visibility.) 'There's no doubt that these campaign managers are savvy to that and it can help get their messages spread.'

The same effect is visible in the UK on Twitter. At the end of November 2019, in the run-up to the UK's third general election in four years, a man stabbed and killed two people on London Bridge in a terror attack. It transpired that he had previously been convicted of serious terrorist offences in 2012 – when a Conservative-led coalition government was in office – and yet released six years later without a parole board assessment. 'Why?' asked the opposition Labour candidate Yvette Cooper.

Priti Patel, the Home Office minister and Tory parliamentary candidate, tweeted her response: 'Because legislation brought in by your government in 2008 meant that dangerous terrorists had to automatically be released after half of their jail term. Conservatives changed the law in 2012 to end your automatic release policy but Khan [the killer] was convicted before this.'[13]

Patel was wrong: the courts had many types of sentence available that could have prevented automatic release. And Khan had been given a sentence in April 2013 (under the coalition), following the passage of the 2012 law change, which confirmed release halfway through his sentence.

But being wrong made no difference. Patel's statement sat on Twitter, attracting thousands of retweets and an entire ecosystem of angry responses. There was no way to kill the untruth.

The palpable anger around the tweet was run-of-the-mill for Twitter, even political Twitter. But compared to the past, it was different. Twenty years earlier, such a claim – made on radio, TV or in a newspaper quote – could have been examined by someone familiar with the law; Patel might have been more closely scrutinised and asked to justify her claims, since they didn't add up. There would have been people blocking the path between her and the public if she wanted to make such a claim.

On Twitter, it could go out to her 120,000 followers, and spread far more widely as it was applauded or criticised. Some were appalled. 'This is an out and out lie,' tweeted Tony Kent, a barrister and thriller writer.[14] 'How is the home secretary being allowed to post this!?! It just is not true. At all. There were plenty of sentencing options available to the Court that would have prevented automatic release. Just. Stop. Lying. You're not Trump. This is not America.'

How was the home secretary allowed to post it? Because it's Twitter, and as long as you don't threaten specific groups you're good to go. (And even threats can be forgiven if you're a sufficiently high-profile political leader, Twitter had previously explained.)

For politicians, 'engagement' is the name of the game. Those thick-skinned enough to ignore the insults could build their following by being controversial and painting complex subjects as binary choices. Reinforcing a political base through Twitter had another benefit: the ideal voter is the one who actually votes, not one who just agrees with you. Polarising content on Twitter could lead to more votes for a politician who could just hit the right, outrageous, note.

As Vaizey observes, 'There's something about Twitter that encourages people to say, "I think you're a dick," when they wouldn't in normal life.'

Brady agrees, yet thinks it can work to politicians' advantage: 'There's this kind of ironic effect where politicians can gain more exposure even if they're getting attacked [on social media] . . . there's a lot of narrative about this in the media about "what is the role of media in getting a president elected?" That's not something I've empirically looked at, but it is a generally interesting question. Even if it's a small role, there's no doubt social media has the potential to play some role, just in terms of exposure. The other thing to consider is that other media tend to cover social media. So even if someone's not on social media, you see people covering, say, Trump's tweets in the evening news, and so there's a way in which things that he's saying are getting more and more exposure.'

We were speaking before the 2020 US presidential election, and Brady was fascinated to know what the four years since 2016 would have taught both sides. 'In 2016, Clinton was definitely less divisive in her messages, but I think there is a little bit of that changing. One of the key things to keep in mind is that politicians are starting to change their strategies.' Viewing Joe Biden's tweets and Facebook postings, Brady observed that 'It looks like Democrats opted to use the same strategy as in 2016, and hope it works this time due to Trump hate.' (Indeed, that was precisely the strategy that Biden's team did use, though the scale of his victory makes it hard to determine how important social media, rather than external events, were to the result.)

★ ★ ★

The big problem with having politicians constantly available on social networks, with their constant churn of ever-new, ever-important updates, is that it can distort the ordinary voter's view of how easily change can be achieved. 'Populism is the natural condition of democratic politics in the age of Twitter,' wrote David Runciman, professor of politics at Cambridge University, in his book *How Democracy Ends*. 'Comments sites encourage instant gratification, when democracy presupposes a capacity for frustration and patience.'

The big danger, Runciman told me, is that the constant ability to question everything, to hold every little piece of electioneering or politicking up to the light and ask whether every little statement is *really* true, and to demand that *something must be done*, has created a sense that every result can and should be questioned.

'Something that has gone is people being willing to accept defeat,' Runciman says. The idea embedded in democratic systems of 'losers' consent' – that those who don't get their way accede to the result, because everyone had their chance – has begun to break down. Previously, he explains, they would think that 'if I hang in with the system, the system will come round to my point of view. The democratic vote would be lost, so we lost this election, but just wait until next time.' He pauses and looks around. 'That has really started to go. Part of what's driving this is this feeling that you can't afford to lose the next election, because our politics is so divided that if you let the other lot in, they will use their power to skew it for next time. There's that breakdown of trust. And a referendum doesn't help, because with a referendum you don't know that there *is* a next time.'

Similar anxieties fuel responses in the US, he says. 'If you look at America, there's deep fear driving the partisanship of politics.' The thinking of the potential losers, he explains, is 'the president gets to pick the people in the Supreme Court, and we cannot afford to be good losers, because while we're being good losers, the president can skew American politics for a generation.' (Appointed for life, US Supreme Court judges have an average tenure of around twenty-five years.) 'Similarly with the [Brexit] referendum – it was presented as a once-in-a-generation vote, and so for the losers, that's really hard to accept.

'And then this technology amplifies those anxieties, and amplifies the possibility of organising around those themes. So I think all [electoral] systems potentially have a weakness, where this technology can press on people's reluctance to wait it out.'

A key element fuelling that reluctance is the impatience that social media, and the smartphone in our hands, generates. Twitter is constant. Facebook always has something fresh. Look, Instagram has new posts! But elections seem as distant as the horizon. 'Elections are years away,' says Runciman. 'But [there's] also that feeling that while you're waiting it out, the world's going to change.' People used to 'switch off' from politics between elections, he suggests; now, there's little respite. 'Three or four years is a long time, and people sense that in those three or four years the whole thing could be skewed against them, and that's amplified by this skewing of communication across these [political] divides.'

Renée DiResta, who researches the use of disinformation on social networks, noted in a September 2016 essay that social media has fundamentally altered the way we think of crowd behaviour, turning something that used to be 'sporadic, spontaneous and transient' into a different beast altogether.[15] 'The most important thing about digitally transformed crowds is this: unlike IRL [in real life] crowds, they can persist indefinitely. And this changes everything.' Online, you can always find a group that can be called to arms by appealing to their common interest: 'You don't have to yell or march or carry a sign, just click the Share or Retweet button from your couch.'

The outcome, she said, would be a battle of radicalised positions: 'The persistent digital crowd is still ultimately a crowd, driven by emotion rather than rational thought ... And since public conversations now happen on social platforms, we're likely to see the polarised, passionate, organised ends of the political increasingly dominate the conversation.'

The last element to consider is voters' interaction with each other over politics, and the democratic process. Here, Facebook in particular has tried to show that it can get people to engage more at the ballot box. Its first attempt to show that social networks really could make a difference to democracy was in 2010, though the experimenters didn't reveal it until two years later. In a letter to the prestigious science journal *Nature* in September 2012, a seven-strong list of authors, including political scientists and, crucially, members of Facebook's data science team, described what they called 'a 61-million-person experiment in social influence and political mobilisation'.[16] The question they were trying to answer was: could Facebook make people more likely to vote? Previous research had already shown that within households, voting can be 'contagious' – if you vote, your housemate is more likely to, and vice versa. Emails urging people to vote, by contrast, were less effective.

On the day of the November 2010 midterm elections in the US, sixty million people found their News Feed had a message encouraging them to vote, a guide to local polling places, and a button to click if they had voted. Some saw a counter of how many Facebook users had clicked the button plus six randomly chosen pictures of Friends who had done so too – a 'social' message. A smaller group of about half a million people were shown the message, the counter and button, but no pictures of Friends; a control group, also about half a million people, saw no message, counter or pictures. By examining publicly available voting records, the experimenters could figure out if they had made a difference.

The effect was real: the 'social' message led directly to about 60,000 people voting, and another 280,000 who voted because of the behavioural 'nudge' of seeing their Friends' messages of having done so. The effect was strongest among close friends; for those outside the ten closest, it wasn't measurable.

A total of 340,000 votes out of 61 million people, or about 0.55 percent, might not sound a lot. But it's plenty, in the right context.

Though Mark Zuckerberg initially dismissed claims that fake news on Facebook affected how people voted in the 2016 US presidential election, you won't find him being similarly dismissive about political advertising. Equally, you won't find him bragging about political advertising having swung an election. That's a tightrope he won't cross, because to deny that advertising has an effect would undermine the company's business model; yet to say that political advertising is effective would lead to calls for regulation of its content and closer oversight.

What isn't widely understood about Facebook's political advertising is the radical difference from that on Twitter (which has since banned them) and Google or YouTube. On Facebook, an advert that 'engages' those who see it is cheaper. This applies to political ads too: a political ad that lots of people 'engage' with, whether by Liking it, sharing it or commenting on it – favourably or unfavourably – will, for the same charge, be shown to more people than one to which people are indifferent. Given all we know about what sort of content propagates best on social networks, that means that outraging, polarising content will benefit, and that a candidate who espouses such views will be more successful in spreading their message. In short, Facebook's advertising model amplifies and encourages populism.

That, allied to the potential for micro-targeting political ads – identifying a small but important demographic who you do or don't want to vote – changes the landscape of political persuasion. In the US, where buying airtime for political ads is allowed, there is a longstanding rule that requires different candidates to be charged the same amount per minute. (Given the perceived importance of TV advertising, this is why candidates amass gigantic war chests.) But on Facebook, different candidates will pay differently according to how engaging the algorithm judges their ads to be. Though Mark Zuckerberg is insistent that political adverts let candidates of all levels reach voters – 'Banning political ads favours incumbents and whoever the media chooses to cover,' he said in October 2019 – he doesn't mention the fact that black-box algorithms decide exactly how much of those challengers you'll see.[17]

In the UK, the starting gun for the Brexit referendum was fired in February 2016. The disparity in attitude soon became clear. The 'Leave' side relied on motivating people who had perhaps never voted to do so, using Facebook adverts as the most effective method first to outrage them about the effects of EU membership, and then to get them to act.

Vote Leave, the main campaigning group, spent nearly £3 million on Facebook ads. Many reached people who hadn't voted for decades, or ever. In the event, 2.8 million such people – who had ignored multiple general and other elections – did turn out, pushing voter participation to a dramatic high of 72 percent.[18] Almost all of them voted to leave, creating a 1.3 million vote margin in its favour. Around a third only decided which way to vote in the week of the referendum, including one in ten who decided only on the day.

The aftermath was full of accusations of manipulation and overspending (the UK imposes strict spending rules on campaigns). One of the biggest questions was about who the Vote Leave campaign had targeted, and what had been said – because Facebook offered no method then of looking at who had bought politically motivated ads. The company finally released the 1,433 Vote Leave ads more than two years after the poll.[19] They included blatant untruths, such as one suggesting Turkey was joining the EU (which would worry those concerned about immigration and Muslims – as Leave-tending voters often did).[20] They targeted a huge range of prejudices and responses: animal lovers (an image of a bullfight), environmentalists (a picture of a polar bear and the false claim that 'The EU blocks our ability to speak out and PROTECT polar bears'), tea-drinkers ('the European Union wants to kill our cuppa', improbably) and many others, using contact data pulled in from a Facebook 'competition' intended to find people who normally didn't vote, which offered £50 million to anyone who could correctly forecast the result of all fifty-one matches in that summer's European Football Championships. (The odds of doing so were calculated at one in five thousand billion billion.)

Facebook said the ads were viewed more than 169 million times, among a population of 66 million. Some people, though, would never have seen any. Others would have seen a lot. Micro-targeting was a big part of the success, driven by data-hungry companies such as AggregateIQ and Cambridge Analytica.

Facebook's precise role in the success of the Brexit campaign is unknown. Nobody *knows* whether the adverts and the News Feed arguments led to the Brexit result. But as with any political advertising, it's naive to think that they had *no* effect.

David Cameron was shocked by the Brexit result. Yet he really shouldn't have been. The Conservative electoral victory in 2015 had been delivered by the careful use of Google ads and micro-targeted Facebook ads in the south-west of England, where the Lib Dems held a number of seats.[21] The Conservatives had aimed adverts at voters there emphasising how the Lib Dems had failed to uphold their principles and instead yielded to the major party in the coalition. The Lib Dem vote collapsed, particularly in the south-west. Among those who were targeted and lost their seat was the Lib Dem leader, who while in government had had to perform a gigantic U-turn on a manifesto pledge not to raise fees for university students. Still, that victim of Facebook's micro-targeting capability was able to make good eventually: three years later, Nick Clegg got a job as Facebook's global head of communications, responsible among other things for defending its policy on political advertising.

The 2016 US presidential election feels now like ancient history. But it marked a significant moment in the demonstration of social warming's effects on politics. Donald Trump won by tiny margins in three critical states. In Michigan, he had 10,704 more votes than Clinton, a difference of 0.22 percent of the votes cast there. In Wisconsin, it was 22,748 votes, or 0.79 percent of those cast. And in Pennsylvania, it was 44,292, or 0.71 percent. In all, fewer than 78,000 votes. Those three states tipped the electoral college vote to Trump. The difference as a percentage of the nearly 14 million total votes cast in all three states: 0.56 percent.

All three states had shown a noticeable trend since Barack Obama's 2008 win: the Democratic winning margin, as a percentage of votes cast in each of the three states, had halved. Gradual demographic shifts towards less-educated white voters meant Clinton was trying to stop a stone rolling down a hill. There were other subtle differences about 2016 compared to the previous two elections: a significant rise in votes going to 'alternative' candidates, and a fall in votes for Clinton compared to Obama's 2012 candidacy. Even so, in Michigan and Pennsylvania she polled more votes in 2016 than Mitt Romney, 2012's losing Republican candidate – which indicates that Trump's campaign roused voters who had not backed the Republican candidate in the past two elections.

So there were multiple effects at play: lower turnout, other candidates – and an insurgent campaign using social media in a

US election in a way it had never been used before, with Facebook use at its highest level ever.

'Facebook and Twitter were the reason we won this thing,' Brad Parscale, Trump's 2016 digital campaign manager, told *Wired* magazine in an interview less than a week after the result.[22] A key reason was feedback: Gary Coby, advertising director for the Republican National Committee, sounded almost awed as he observed that 'Their platform's built to inform you about what people like and dislike.'

Speaking in December 2016, Parscale said his analysis by the last week before the November election suggested that the likeliest states that Trump could win were Wisconsin, Michigan, Pennsylvania, Ohio, Florida, and five others that had been among the closest, with vote margins of less than 10 percent, in 2012. The Trump campaign spent around $100m on digital messaging in the last weeks ahead of voting day, and $5m in targeted digital advertising in those three states.[23] 'So I was moving targeted money around,' Parscale told NPR in December 2016.[24] He felt that the immediacy of his effort had been key to the result: 'I think you see that – how we won Wisconsin and Pennsylvania and Michigan. If I would have waited one more day [before acting], maybe that had the effect of the, you know, 100,000 votes.'

There was also a notable difference in the advertising media chosen by the Trump and Clinton campaigns. Parscale allocated $39m on final-days TV ads, and $29m on digital ads; Clinton's team spent $72m on TV ads, but just $16m on internet ads. During the election the Trump campaign ran literally thousands of different adverts every day, using slight variations and testing responses. On one day, the number hit 175,000. Most were to encourage people to vote, Parscale says. But some were meant to discourage: 'We would target those people who we felt were in the middle till we could move them over either to an "undecided" or back into a Trump column.' Potential Trump supporters were found by uploading the names of existing supporters, and then getting Facebook to find people with similar interests (its 'Custom Audience' tool), and then reaching them through targeted ads. The numbers targeted could be tiny – 100 or so – but the power of similarity meant Parscale was increasingly confident of results. Even when Trump harangued him for not spending as much on TV ads, he remained certain, telling Trump to his face that 'If you are going to be the next president, you're going to win it on Facebook.'[25]

Parscale had been telling Trump's campaign this same message as far back as July 2015. The social network's expansion beyond its early years of mainly urban users, to one where it was essential for rural dwellers too, made it perfect for reaching people who might not otherwise vote – the same sort of audience that Vote Leave had targeted.

Trump's campaign wasn't the only one trying to inflame passions. Russia's Internet Research Agency (IRA) had hundreds of people employed full-time generating and posting content on social networks secretly intended to galvanise certain sections of the American public and discourage others. The Mueller report, which included the most thorough investigation of what had happened online, found that the IRA's posts, comments and tweets had reached at least 29 million, and perhaps as many as 126 million Americans on Facebook.[26] The plan was no overnight scheme; work had started creating the accounts in 2014, with the clear long-term aim of using them to inflame the antagonism that was sure to be a feature of the 2016 election. They spread into YouTube and Twitter, and later to Instagram and Tumblr. Some of the Facebook Pages they created were enormous by any standards: 'United Muslims of America' was the largest with 300,000 followers, followed by 'Don't Shoot Us' with 250,000. On the other side of the political divide, the 'Being Patriotic' and 'Secured Borders' groups had more than 200,000 and 130,000 followers respectively. The agency covertly ran two opposing networks that managed to persuade real Americans to turn out to protest against each other at rallies in Miami and Philadelphia. But sometimes the purpose was just like Parscale's: to suppress votes. Some IRA accounts pretended to be black activists disdainful of Clinton and sought to discourage black voters from supporting her.

Were those 78,000 votes in those three states down to Parscale? The IRA? Profiteering Macedonians writing fake news that tended to be pro-Trump (or anti-Clinton) because that garnered more clicks? In this hyper-connected world, where one presidential candidate was using his social media presence to wrest attention from rivals and set the agenda, and the biggest social networks were seeing more and more partisan content, to say there was no effect seems nonsensical. As with advertising, if it doesn't affect what people think, why do it?

The famous quote about advertising used to be that 'half of what I spend is wasted – I just don't know which half.' Online advertising was meant to change all that: people would be more precisely targeted with things they were known to be interested in, and advertisers would only pay for the advertising that people acted on. That, after all, is how Google works: entering text into the search box reveals some intent about what you want to do or find out. Google auctions off space in the search results page to the best bidder who can satisfy that, getting paid only if someone clicks the link. That auction, repeated billions of times every day, hauls in billions of dollars every year, even though on average only one in every forty search ads displayed gets clicked.[27]

The usual objection to discussion about the effects of social networks (and thus social warming) on politics is that they don't *change* anything; that instead they strengthen what was already there. But that's the same as saying an advertiser can't make you buy something you don't want. What social warming means with politics is that interactions are more heated; positions can quickly become more entrenched.

As the 2020 US presidential election came into sight, the contradictions of Facebook's policies over political advertising were subjected to close scrutiny. In October 2019, Mark Zuckerberg was quizzed in the US Congress by Alexandria Ocasio-Cortez: the young billionaire facing the youngest member. Ocasio-Cortez quickly impaled Zuckerberg on the horns of his self-made dilemma: even if a statement had already been shown to be false by Facebook's fact-checkers, it would still be allowed in a political advert.[28]

'You announced recently that the official policy of Facebook now allows politicians to pay to spread disinformation, in 2020 elections and in the future,' she began. 'So I just want to know how far I can push this in the next year. Under your policy . . . could I pay to target predominantly black zip codes [addresses] and advertise them the incorrect election date?'

'No, Congresswoman, you couldn't,' Zuckerberg responded. 'We have, even for these policies around the newsworthiness of content that politicians say, and the general principle that I believe—'

'But you said you're not going to fact-check my ads?'

'If anyone, including a politician, is saying things that can cause, that is, calling for violence, or could risk imminent physical harm, or voter or census suppression, when we roll out the census suppression policy, we will take that content down,' Zuckerberg said. He seemed like a diver coming up for air as he recalled that this was actually covered by a policy.

'So you will— There is some threshold where you will fact-check political advertisements. Is that what you're telling me?'

'Congresswoman, yes, for specific things like that where there's imminent risk of harm.'

'Would I be able to run advertisements on Facebook targeting Republicans in primaries, saying that they voted for the Green New Deal [an expensive scheme to tackle climate change]? I mean, if you're not fact-checking political advertisements— I'm just trying to understand the bounds here of what's fair game.'

Zuckerberg wasn't sure: 'I think, probably?'

'Do you see a potential problem here, with a complete lack of fact-checking on political advertisements?'

'Well, Congresswoman, I think lying is bad, and I think if you were to run an ad that had a lie, that would be bad. That's different from it being, from in our position, the right thing to do to prevent your constituents . . . from seeing that you had lied—'

'So you won't take down lies, or you will take down lies?'

'In a democracy, I believe that people should be able to see for themselves what politicians, that they may or may not vote for, are saying, and judge their character for themselves.'

'You may flag that it's wrong, but you won't take it down.'

'It depends on the context . . .' Zuckerberg trailed off.

In a follow-up question, Sean Casten, another Democrat, asked whether a member of the American Nazi Party (which two years earlier had been banned from Twitter for 'using or celebrating violence to achieve their aims') could say things in a political advert while running for office that would be deleted if they were not.[29]

'That depends on a bunch of specifics . . .' Zuckerberg answered.

'Wow,' retorted Casten. 'I don't think that's a hard question.'

In fact, Zuckerberg almost certainly had mulled over that question before. In 2015, senior executives at Facebook agonised over the question of whether to ban Donald Trump's campaign video in which he suggested stopping any Muslims from entering the US – based on nothing but their religious belief. As reported by the

Washington Post, and effectively confirmed by Facebook, they drew up a list of four options: remove the video, allow it under a one-time exception, create a carve-out from their rules for 'political' speech, or effectively throw the Facebook guidelines on what counted as hate speech out of the window and allow anything.[30]

Monika Bickert, who as head of public policy often speaks to the media to offer Facebook's non-apologies for all the ills that transpire on the platform, drew up a list of possible PR impacts. Among them was whether lowering (or abandoning) standards or creating political carve-outs implied that Facebook would have provided a platform for Hitler, had it existed in the 1930s – a dilemma that was dubbed the 'Hitler question'.

Partly at the urging of the former Republican operative Joel Kaplan, then effectively the No. 3 in the company, the Facebook executives decided to create a political carve-out. Lying was OK; there was no need to create an exception for politicians there. But politicians would be allowed to use hate speech that would be removed if posted by anyone else. Instead of giving everyone an equal voice, Facebook gave a particular group – politicians – special status.

A month after Zuckerberg was quizzed by Ocasio-Cortez and Casten, the satirist Sacha Baron Cohen – best known for his Ali G character, a faux-idiot – gave the keynote address to the Anti-Defamation League, an independent group that opposes anti-Semitism and extremism.[31] 'Under this twisted logic, if Facebook were around in the 1930s, it would have allowed Hitler to post thirty-second ads on his "solution" to the "Jewish problem",' Cohen said, in a lacerating speech that didn't spare any of the big social platforms. 'We're not asking these companies to determine the boundaries of free speech across society. We just want them to be responsible on their platforms. If a neo-Nazi comes goose-stepping into a restaurant and starts threatening other customers and saying he wants kill Jews, would the owner of the restaurant be required to serve him an elegant eight-course meal? Of course not! The restaurant owner has every legal right and a moral obligation to kick the Nazi out, and so do these internet companies.'

But Zuckerberg shows no inclination to do that. The answer to the 'Hitler question' appears to be that *of course* a 1930s Facebook would have allowed the ads. It already does for politicians in the modern day who incite racial and ethnic hatred. In January 2020, Brazil's president Jair Bolsonaro suggested in a Facebook broadcast that the indigenous peoples in the country were 'increasingly becoming human beings just like us'. The broadcast was allowed to remain online. It wasn't a part of a political campaign. What exception was at work for allowing it? In India, the country with the largest number of Facebook users, a local executive personally intervened to prevent a ban on a ruling Hindu nationalist party politician whose posts had been flagged by moderators for hate speech: he had called for Muslims to be shot and mosques to be destroyed. The executive told staff that banning the politician could hurt Facebook's business prospects.

In the US, one Facebook executive went even further in the tacit acceptance of rabble-rousing on the platform. 'Right-wing populism is always more engaging,' the executive told the American politics news site Politico in September 2020.[32] The aim was to deny the idea that Facebook's algorithms favour conservative American views (which consistently showed up as those with the most 'engagement' on the site). Such content, the executive said, appealed to the most primitive emotions by pushing topics such as 'nation, protection, the "other", anger, fear'. And: 'that was there in the 1930s. That's not invented by social media – you just see those reflexes mirrored in social media.' Such an analysis overlooks the existence of successful populist socialist movements such as Hugo Chávez in Venezuela, Fidel Castro in Cuba, Syriza in Greece, Podemos in Spain, or even the intense support among many electors for Jeremy Corbyn's left-wing policies in the UK. As a historical analysis, it was somewhat wanting. A spokesman for Joe Biden's campaign was withering about the comments, and focused on what he saw as the real source of the problem: 'It is a choice to create an algorithm that feeds the distrust and polarisation that are tearing us apart.'

Is Facebook content to stand by while politics becomes more extreme through the inherent dynamics of its system, while politicians lie in adverts to their voters, while those in power use its platform to cement their position? It's not a hard question.

Yet the question about influence lingers. A study carried out in late 2017 and published in the prestigious *Proceedings of the National Academy of Sciences* (PNAS) by a team at North Carolina's Duke University and the University of Copenhagen seemed to suggest that the Russian Twitter efforts had had no impact.[33] The study looked at the political viewpoints of 1,239 voters, both Republican and Democrat, who had interacted with some of the IRA accounts by talking to them or liking or retweeting their content.

The first research finding was that the people most likely to interact with such accounts were either strongly interested in politics, or else heavy users of Twitter. (Both, when you think about them, are predictable.) The second was that over the course of a month the IRA's tweets didn't make any difference to the political views of anyone, as measured on a five-point scale of ideology.

However, the authors hedged their zero outcome with many caveats about what they couldn't be certain of. *Possibly* trolls affected independent voters, or those who weren't very involved in politics, more strongly. (In the US, independent voters often determine elections because they haven't indicated loyalty to either party.) *Perhaps* they changed society's view of certain issues, or of a politician's trustworthiness. 'We also could not study whether or not troll interaction shaped voting behaviour,' the researchers added, although their study would be reported as 'Russian Trolls Aren't Actually Persuading Americans on Twitter, Study Finds'.[34] Certainly, in the study's findings the trolls weren't persuading that group of Americans post-election. But that is still some distance from having no impact. The timescale was short, the cohort wasn't likely to change views, and – crucially – the election had already happened; the study couldn't find out whether the efforts had made a difference in the past.

The problem with believing studies that say that social media has no effect, especially in the face of the Brexit and Trump votes (and, as we'll see in a later chapter, voting in the Philippines earlier in 2016, and subsequently in Brazil and other countries), is that studies of the effects on Americans of election efforts imply *nothing* works. A 2017 paper from the University of California at Berkeley performed a meta-analysis of multiple studies looking at the effects of actual contact by political campaigns with voters.[35] The conclusion: 'The best estimate for the persuasive effects of campaign contact and advertising – such as mail, phone calls and canvassing – on Americans' candidate choices in general elections is zero.' The conclusions for online and TV advertising were the same, though researchers allowed that there was 'less evidence'.

So: two studies showing zero effect. The latter one in particular seems to imply that the billions of dollars spent on political advertising and canvassing is utterly wasted, because nobody changes their mind.

However, the Berkeley researchers glossed over some crucial points. While the *average* effect might have been zero, in two of the six states they examined – specifically Ohio and North Carolina – the canvassing did find 'subgroups of voters . . . more likely to be persuadable' and that the canvassers 'ultimately had some persuasive effect targeting these voters'. Nor did they take into account that many of the trolls targeted a specific group – African Americans – in an effort to suppress their votes.[36] And 2016 was the first time since 1996 that African American voter participation fell, by seven percentage points to 59 percent. Not all of that could have been the effect of IRA trolling. But it seems improbable to suggest that *none* was.

Politicians have noticed the extent of social warming, and particularly its effect on their profession. In October 2020, the Democratic senator from Virginia, Mark Warner, wrote separately to Facebook,[37] Google[38] and Twitter[39] to scold them for their lack of action over misinformation and the algorithmic amplification of untruths related to the election. He called Facebook's and Twitter's labelling of misleading content, particularly Donald Trump's comments on the trustworthiness of postal voting, 'wholly inadequate', and said that YouTube was a major source of misinformation on the topic, spread by right-wing bloggers. WhatsApp, he noted, was being used to spread election misinformation in Florida: 'What are you doing to address this growing threat WhatsApp poses in the United States given the ways in which it has facilitated violence and electoral turmoil in other countries?' Warner demanded of Zuckerberg, and pointed out to all three companies that the Pew Research Group had reported in September 2020 that three-quarters of Americans had 'little or no' confidence that the big platforms could fight misuse in the election.

Warner's complaints focused around the implementation of the 'Honest Ads Act', a piece of legislation that had been floating around the US Congress since 2018 but wasn't actually law. Facebook had been quietly resisting the act's passage, telling lawmakers that it was already complying with the act's requirements – labelling ads, creating a repository of them – and so there was no need for an actual law. (The nice thing about self-regulation, after all, is that the penalties are so much lighter.)

He also encouraged all three platforms to introduce the equivalent of 'bug bounties' – payments to hackers who tell companies about flaws in their software rather than exploiting them – for researchers and journalists who identify violations of the rules.

The fact that Facebook, Twitter and Google all made changes to their policies around political advertising ahead of the US 2020 election shows that they are acutely aware of their potential effects. Twitter banned it altogether; Google stopped any attempts to micro-target small groups of users in the way that Facebook enables. And Facebook itself introduced country-of-origin checks and an Ad Library that could show who had advertised what, and roughly to whom – though not the fine detail of why and how people were targeted. Some light began to fall on the dark ads that had marked previous elections everywhere.

The message is not that persuasion doesn't work. It's that the effect is spotty and hard to pick out from the noise of indifference. If nobody ever changed their mind about a political candidate or politics, we wouldn't see the changes in governance that we do.

Pointing to a particular person and saying that their opinion and hence vote was swayed by this advert or that piece of viral content is impossible, and anyway isn't how the process works. Social warming makes tipping points in finely balanced situations more prone to occur, and pushes the extremes further away from each other. In politics, where national borders and the models of rule remain the same, that means people are confronted more and more with views they disagree with, espoused by politicians they dislike – who can flourish by appealing to more and more extreme versions of their party's politics. As long as the reward for being outspoken is more attention, especially on social media, the cycle will continue. And as long as the networks value revenue above any duty to prevent politicians inspiring hate, the situation will get worse.

7

FACT TO FAKE: HOW THE MEDIA ECOSYSTEM COLLAPSED

By and large, Twitter is reserved for people to act like bigger idiots than they would otherwise. It is easy to be racist or sexist or generally asinine in 140 characters. It is impossible to be sublime. – Callbuzz, April 2012

At about 3.30p.m. in New York on 15 January 2009, Janis Krums took a photo on his phone that changed the trajectory of world media.

Krums wasn't a journalist. Aged twenty-three, he had an iPhone with which he would occasionally post tweets to his 170 Twitter followers. He enjoyed photography, and would use a service called Twitpic that allowed him to link photos to his tweets.

As Krums stood on the commuter ferry that Thursday afternoon, an Airbus A320 passenger plane ditched in the river: both engines had been knocked out by geese soon after its take-off from the nearby LaGuardia airport. He snapped a picture and uploaded it to Twitpic and added a tweet: 'There's a plane in the Hudson. I'm on the ferry going to pick up the people. Crazy.'[1]

There was no 'retweet' function in 2009, but people copied the link to the photo, and took their own screenshots of the tweet, and spread it far and wide. Inside Twitter's offices, staff gathered around a TV screen showing the photograph, which was going viral.[2] Krums's tweet echoed around the world, and suddenly made clear that Twitter, then less than three years old, had the power on its own to transmit news events to millions of people – and for free. Twitter users broke the news about fifteen minutes ahead of the normal media that day. 'I posted it because I thought "that's pretty newsworthy" and I wanted to share it with the people who follow me on Twitter,' Krums said later.[3]

That incident 'changed everything', recalled Jack Dorsey, the Twitter co-founder, a couple of years later. 'Suddenly the world turned its attention because *we* were the source of the news, and it wasn't even us, it was this person in the boat using the service. Which is even more amazing.'

Twitter's importance as a news medium was reinforced in June 2009, following a disputed election in Iran. Protesters on the streets were met with violence from the police and paramilitaries. Many also took to Twitter to express their anger. The US State Department viewed it as so important to keep their voices heard that it tasked Twitter to delay an upgrade that would have taken the service offline.[4] But there was also misinformation and confusion: some got the mistaken impression that all of Iran was against the result. 'You trade speed for accuracy,' conceded Clay Shirky, discussing the conflicting reports that were emerging, some of which couldn't even be verified as being in-country.[5] But mistakes could quickly be corrected, and misinformation would be evident, he added. The commentator Andrew Keen disagreed: Twitter, he said, was terrific for distributing opinion, but in a crisis the tweets from the street couldn't take the place of proper, careful media coverage. At the *Guardian*, the reporter Matthew Weaver, who ran a liveblog from the London office as the events unfurled, disagreed with Keen: when something happened, the tweets would show up first, then pictures, then video, then reports from journalists. 'What people are saying at one point in the day is then confirmed by more conventional sources four or five hours later,' he explained.

Twitter soon became accepted as the medium through which we could get closer to the truth of events. For fast-moving events in particular, it became a way for journalists to get little snippets of information, or a running commentary, out to an audience that increasingly wanted to know *right now* what was happening.

This was great business for Twitter. For news organisations, the benefits would turn out to be a lot more mixed.

Four years earlier, I had been sitting in a huge hall in the grounds of Windsor Castle, one of a group listening to speaker after speaker try, and mostly fail, to come up with an overarching definition of 'news' in the age of the internet. We were a motley collection of journalists, would-be politicians and businesspeople who had been invited to spend a couple of days forecasting the future of news: what it would look like, how it would work, who would want it, whether there would still be a business in it. At the time, I had just started working at the *Guardian* as its technology editor.

What the hell *was* news, I wondered. To open the conference, a few of the attendees spoke briefly about how they viewed news, and how they got it, and what they thought it was. 'Important events . . . the stuff somebody doesn't want you to print . . . things that you need to know . . .' The familiar phrases floated by, but nothing we hadn't heard before. Most of all, they didn't answer the question of why fewer and fewer people wanted to get this magical product from physical newspapers. Mass media was about the masses, and often showed a patrician view of what was good for you, in an eat-your-greens way: of *course* you should be interested in foreign policy; of *course* you should care about the effects of climate change on people far away. Of *course* you should care about the maximum betting limit of fixed-odds betting terminals in gambling dens. But do you?

Although online readership was ramping up, revenues were not compensating for the falling income from print. Nobody mentioned how mysterious it was that Google, which had gone public the year before, was making money hand over fist from selling ads online.

Then another speaker came to the front. She wasn't a journalist, she explained apologetically, but she'd do what she could. She began describing her morning routine for finding news, which involved going to her computer, opening a few bookmarked web pages for topics she was interested in (computing, transatlantic news, a couple of other things), and working through them. 'I'm looking for . . . for, well, stuff I care about, stuff I want to pass on,' she said, slightly at a loss.

For me, the phrase was like a lightning bolt. After so many years trying to find a definition of news, here it was on a plate. *Stuff I care about. Stuff I want to pass on.* It's so simple. Some news is both (you won the lottery, perhaps?); some is just one. A change in tax rules that affects your income is probably 'care about', while a cute puppy doing something silly is hardly important, but fun to share: *pass it on.*

The problem that our conference couldn't quite see at that point, because Facebook was still limited to Harvard students, and Twitter hadn't been invented, was how social networks would take both ideas and turbocharge their effectiveness – and that news organisations would be left floundering in their wake because they would cope so badly with the individual nature of 'stuff I care about', while 'stuff I want to pass on' would become plentiful, and tuned not to the accuracy that news organisations care about, but to how eager people would be to share it – a process that would become easier and easier and easier.

The effects of this would take some time to become apparent – and involve news organisations sabotaging confidence in their own brands, capricious changes to Facebook's News Feed algorithm and journalists coming into open conflict with managers – concluding with distrust just when the opposite was most needed.

Publishers in the age of print, whether local, regional or national, all worked to a similar formula: package up lots of different content, some of which people would be interested in and a lot they might not, sell some adverts around it, and hope people wouldn't find any better way to distract themselves on their morning commute, or in their spare minutes. Individualising content to create the 'Daily Me' was an idea often brought up in chin-stroking meetings about the future of publishing online, because the idea of delivering articles or programmes that people specifically wanted seemed vaguely possible with the online world. The reality, courtesy of Facebook and Twitter, turned out to be far less comfortable than they had expected.

What few grasped was how the internet would change the concept of news, and its inherent value – which is close to zero. In 1996, Shirky predicted the problems that traditional news organisations, would face. Consumer use of the internet had barely begun, yet he foresaw a colossal change coming that would disrupt the news business in the way that a runaway locomotive disrupts buffers.

In a blogpost titled 'Help, the price of information has fallen and it can't get up', he dismissed the suggestion being made by traditional publishers that internet users would 'eventually have to pay for content'.[6] Wrong, he said: the price of information was in freefall, because pretty much anyone could now generate and distribute content. The internet had utterly changed both the supply of content, turning scarcity into surplus, and its distribution, which for many forms of content was effectively free: 'One of the principal effects of the much-touted "Information Economy" is actually to devalue information more swiftly and more fully,' he wrote. Shirky compared the potential rise of independent news sites to the end of the monopoly that the three US TV broadcast networks had enjoyed in the US from the 1950s to the 1980s.[7] They had effectively had a licence to print money, until the arrival of cable networks fractured the audience among hundreds of channels all competing for attention. The internet would do the same, but on an even more precise scale: 'The internet is a massive medium, but it is not a mass medium,' he wrote.

When Shirky wrote his post, the threat of free-for-all information looked remote. In many American cities, newspapers enjoyed effective monopolies that made them highly profitable products.

The first threat to this ecosystem was the arrival of Craigslist, which offered users the chance to list services or products they had for sale or hire, or that they wanted. From its beginnings as an email list in 1995, it quickly blossomed into a web service, initially for San Francisco in 1996 and then in 2000 for other parts of the US. Craigslist offered the internet equivalent of classified adverts – the sort of short ad that people had for decades paid substantial sums to insert into local papers. Craigslist was free; only in 2004 did it start charging for anything, and that was a flat $25 for job listings ($75 in San Francisco).

To thousands of newspapers across the US, Craigslist was an economic termite that slowly ate away at their foundations. In 2004, 36 percent of the $48 billion of US newspaper advertising, or about $17 billion, came from classified ads.[8] A study found that in the same year Craigslist had leached away ads worth $50 million to the papers, but perhaps one-tenth of that to Craigslist. 'The media conglomerates are still extremely profitable,' the site's founder, Craig Newmark, told *Fortune* in 2005.[9] He didn't want to see journalists' jobs hit, but he was sure that 'serious journalism' would still be in demand and well paid in the future. Still, $50 million out of $17 billion didn't seem like a lot to worry about.

Nor did Google's arrival on the stock market in summer 2004, when its total revenues were $3.2 billion. Newspapers were still a far bigger industry.[10]

In reality, the habitat that supported the newspaper industry was beginning to disintegrate. Advertising revenues in the US peaked at $49.4 billion in 2005, and again in 2006, and then collapsed as the combined effects of the financial crash, Google and Craigslist became evident. In 2008, they fell by 17 percent year-on-year to $37.8 billion, and then by nearly 30 percent as the Great Depression hit. Staff were cut just as heavily.

At the same time, social networks took off, and accelerated the competition to existing news organisations from newer publishers. Easily available advertising inventory and blogging templates meant anyone could put up a news website. And because the internet thrives on bringing together people with niche interests, you could run an advertising-supported site about nearly any topic. Social networks were also a great place to attract readers. They essentially flattened the landscape: on Twitter or Facebook, a link to a deeply researched story from the biggest news organisation could sit next to one from a lone blogger with an opinion about the news organisation's piece, and then another with a picture taken from the scene of a breaking news event (such as Krums's). 'Stuff I want to pass on' became the byword for everything.

Were such links an opportunity or a threat to the news organisations? The gigantic, and growing, number of people using social networks offered the chance to snag readers who had never seen a publication and capture their interest. But the competition for attention meant that once-loyal readers on paper could be distracted away on screens by rivals. However, the social networks definitely did bring people who otherwise wouldn't have come to the site. News organisations entered an uneasy symbiosis with social media.

While older news organisations played by the rules that they understood, the new ones made up their own. They saw little point in trying to be the next *New York Times* or *Guardian* – established, monolithic news organisations covering the entire waterfront of content, from general news to politics to sport to fashion to leisure. In the case of Upworthy and Buzzfeed, the secret to success was to write stories intended to go viral, so that hitting the jackpot with enough hits would make up for the misses.

The arrival of Twitter with its hubbub of news, large and small, transformed journalism. From being 'the place where people talk about their breakfast,' Twitter grew to being the place where news could be found, fully- or half-digested. It was also where 'uncensored' tales could be told by non-journalists. Often these were too good to ignore, and so were featured in the news. That feedback loop validated Twitter, making journalists even more eager to be there. And journalists, hating to be beaten to the punch by rivals (or amateurs), would compete to put out the early 'take' on a breaking story. Within a few years, the relationship was unbreakable. 'Once we had a real-time channel of constant, continuous information available to us, we could never do without it again,' observed Matthew Panzarino, a technology journalist who covered its rise at TechCrunch and other sites.[11] 'Even if Twitter the company had gone under, or screwed up so badly that somebody had to step in and do something very similar in a different way, that particular firehose would need to exist.'

Unlike many of the other social networks, Twitter's first implementation was precisely what journalists wanted. For them, there were three crucial advantages over Facebook. Connections didn't have to be mutual, allowing ad-hoc conversations to spring up everywhere; everything was public by default, so any information put on it became usable; and (until 2016) there was no algorithmic meddling with the timeline, so everything was in strictly reverse chronological order with the most recent events first, like a news ticker tape.

The importance of the latter point was demonstrated in August 2014 when protests began in Ferguson, a town in Missouri, over the shooting of an African American man called Michael Brown by a police officer who suspected him of robbery. The unrest continued for days, and was evident on any news source, including Twitter. Yet on Facebook, most of the News Feed was occupied with the feel-good 'ice bucket challenge', observed John McDermott at Digiday.[12] That was because Facebook wasn't about news, but about what your Friends were sharing. 'Relying too heavily on Facebook's algorithmic content streams can result in de facto censorship,' McDermott warned. 'Readers are deprived a say in what they get to see, whereas anything goes on Twitter.'

The new publishers ran incredibly lean operations using eager young staff who would scavenge the internet for news in whatever form they could find it: Facebook posts, celebrity tweets on Twitter, content from better-funded sites. The capitulation to Facebook, Twitter and Instagram was understandable: they were excellent sources of direct quotes from politicians, celebrities and ordinary people – the fodder that had fuelled journalism for centuries. If someone captioned a photo on Instagram, or wrote a tweet, there was no need to ring them up and ask them to say something. Social networks functioned as press release and, often, photocall. With staff numbers at publications falling, the social networks that had once been seen as interesting means of picking up more readers instead became the tail that wagged the dog.

It also became evident that truth and accuracy were not essential to the output of a 'news' organisation. For decades the British and American tabloid press had thrived on publishing stories that often bore only a nodding acquaintance to real-life events. Now they were up against sites willing to write wildly partisan versions of what had happened in order to attract views. After all, 'stuff I care about, stuff I want to pass on' makes no mention of accuracy. The ferocious churn of social networks selected for attention, not truth.

'In 2014, it was still a really big problem,' says Craig Silverman, now the media editor for Buzzfeed News, but at that time working at the Tow Center for Digital Journalism at Columbia University on tracking rumours. 'The incentives were around getting big traffic, having things go viral, getting a quick story that lots of people will read, and will climb up your Chartbeat or Parse.ly dashboard in the newsroom' – referring to analytics packages that show how many views a story is getting. 'These incentives led to a lot of really bad behaviour within native digital newsrooms and legacy newsrooms trying to become digital, certainly in the range of like, you know, from 2010 until about 2018 or so. Anything that was viral and had potential for big traffic, somebody was going to jump on and write about and not necessarily apply strong verification. You had entire news sites that were just built on finding the most viral stuff.'

Some sites, such as Gawker, paid journalists according to how many page views they got – a pressure entirely new to journalists working in text (though familiar to TV executives), who usually get a simple salary with no bonus structure and no output quota.[13] That gave journalists another purely selfish reason for getting noticed on Twitter. Plus if they could show they were influential there, and direct traffic to their organisation, that might translate into leverage in future salary discussions.

Even while journalists delighted in their new-found fountain of news and sources, Twitter exacerbated another effect that was hitting publishers: the atomisation of news. Just as the introduction of digital music downloads had cemented the idea that music didn't come in collections of songs called 'albums', but that each song could be chosen (or ignored) individually, so Twitter picked apart the 'newspaper' or 'bulletin' into its constituent parts. The 'story' became the unit of journalism.

But the news business faced an even bigger challenge than the music business had. When musicians and record labels – the equivalent of journalists and publishers – had seen their lucrative packages of CDs holding multiple songs atomised into individual downloadable tracks, they reoriented themselves around new ways of presenting that same content, in mixes, playlists, and all-you-can-eat monthly deals with companies like Spotify. By 2019, that brought total recorded music revenues in the US to nearly two-thirds of the 1999 peak, when CDs ruled the world, and the biggest for over a decade.[14]

Musicians, however, weren't competing on those music services with recordings of your cousin's new baby, or your friends singing 'Happy Birthday'. By contrast, news organisations competing for visibility on Facebook and Twitter were up against every other potentially attention-grabbing piece of information, including the aforementioned new baby, and everything else your family or friends might have posted. In theory, Facebook's algorithm would push external news sources about topics that you'd like to read into your News Feed. Yet that didn't seem to happen.

Journalists and publications competing for views on Facebook and YouTube made many of the same complaints that musicians made about streaming services: that compared to the past, an incredibly large number of people had to consume your work in order for you to get paid a usable amount; and that they were at the mercy of capricious masters whose whims could deny them visibility from one day to the next with no coherent explanation of what they had done wrong, or what they needed to do to regain it. The media's reliability in the eyes of readers was drastically undermined by two developments, both driven through social networks: the 'pivot to video' encouraged by Facebook, and the rise of 'fake news'.

The 'pivot to video' was a result of media organisations looking to take advantage of Facebook's enormous and growing reach, and the News Feed that increasingly looked like the golden goose – albeit one whose demands for food varied wildly, as did the size of its eggs. A watershed moment came in 2014, when Facebook told publishers and advertisers that a new slew of data showed that what people really wanted to see in their feeds was video. Words, and even still pictures, were old hat.

Facebook pushed this line because Zuckerberg was concerned about the competition from Google's YouTube: the video service was racking up huge amounts of 'viewing time'. But publishers also thought video looked attractive: adverts inserted in video content paid between six and ten times more per 'view' (measured according to an industry standard) than those running alongside text.

On hearing Zuckerberg's call to arms, a number of news publishers decided that any journalist who couldn't be relied on to take part in the 'pivot to video' was surplus, and duly fired them. The remaining (or newly hired) staff were required to make more and more video content. But news-quality video is much more difficult and time-consuming to produce than text and pictures. Short video, which seemed to work best, was even more difficult to execute well than filmed interviews. Snappy editing, quick-fire messages, complex effects – none was as straightforward or quick as finding out some facts, asking some people their opinions, and synthesising that into text to put on a web page.

Given enough people creating content, though, and with an algorithmic leg-up from Facebook's News Feed, surely video would be a big hit. 'We're entering this new golden age of video,' Zuckerberg told Buzzfeed News in April 2016.[15] 'I wouldn't be surprised if you fast-forward five years and most of the content that people see on Facebook and are sharing on a day-to-day basis is video.'

Speaking a couple of months later, Nicola Mendelsohn, Facebook's vice-president for Europe, put some numbers on it: 'We've gone from a billion video views in a day to eight billion views a day, in a year,' she told a Fortune conference in London in June 2016.[16] 'In five years ... it will probably be all video,' she predicted. 'We're seeing a year-over-year decline on text ... if I was having a bet, I would say: video, video, video.' She was actually some way off the pace: Zuckerberg had announced the eight billion figure in November 2015 on a Facebook earnings call, saying that had risen from four billion per day in April.[17]

News publishers who dutifully followed the instructions and concentrated on video were puzzled by one thing: their visitor numbers cratered. Even though Facebook said video was a big hit, the readers – now viewers – seemed to think differently. With fewer people coming to their site, publishers became increasingly reliant on the black boxes of Facebook's News Feed and Google's YouTube. Journalists, meanwhile, began to wonder quite how effective the 'pivot to video' really was.

What they didn't know was that the evangelism of Mendelsohn and Zuckerberg had been based on deeply flawed data. Video was nowhere near as popular as Facebook had made out, as news sites and advertisers began to realise as their visitor numbers kept dropping.

The flaw lay in the way that Facebook counted a 'view'. Videos were set to autoplay in the News Feed, but registered as 'viewed' if they played for three seconds or more – a remarkably short time compared to the industry standard set by YouTube, which defined an ad view as lasting thirty seconds or more. On Facebook, almost anything would count as a 'view' unless you scrolled quickly past. This was quite an exaggeration of people's actual interest.

In September 2016, the fallacy began to unravel. A post on Facebook's 'Advertiser Help Center' noted quietly that for two years or so, the 'average duration of video viewed' had been significantly overestimated.[18] Instead of taking the total time spent watching a video (including the sub-three-second group) and dividing that by how many people it was shown to – including those who scrolled past – the calculation had summed up the total time (including the sub-three-second group), but then divided it *only* by the number of people who 'viewed' the video – the over-three-second group. So if a fifteen-second video was shown to two people, and one watched all of it and the other scrolled away after two seconds, that would count as seventeen seconds of total viewing, by just one viewer – a miscalculated 'average' of seventeen seconds, which is longer than the actual video, instead of eight and a half. The more people who scrolled past before the video played, the bigger the error would be. Paradoxically, if people really hated your video, the miscalculation would suggest it had done really well – a perverse signal that made things worse for publishers.

Facebook carefully didn't say how big the discrepancy was, and insisted that figures given out previously for 'time spent watching video' and the 'number of video views' (those over three seconds) were still correct. Some advertisers were quietly told the 'discrepancy' had overstated the figure by 60 to 80 percent. But others said the gap was far bigger – as much as tenfold, which would imply that only a fraction as many people actually watched – and that they had raised concerns with Facebook in 2015, pointing out that some of the numbers Facebook was reporting made no sense, giving videos average watch times longer than their duration.

Advertisers filed a lawsuit alleging fraud and claimed Facebook had sat on the revelation for a year. Facebook didn't admit wrongdoing, and the case was eventually settled for $40m.[19] Documents released as part of the lawsuit showed that the company had discussed a 'no PR' strategy so as not to bring attention on its faulty mathematics.

Yet as Will Oremus observed in *Slate* (which had gone through its own pivot-to-video/dump-video rinse cycle) in October 2018, while the lawsuit was ongoing, the whole idea was built on a lie.[20] *Users* didn't want more news video. 'It was always about what advertisers wanted: the captive attention of consumers,' he wrote. A separate analysis of 100 million videos found that the topics that generated the most engagement were food, fashion and beauty, animals (particularly pets), and DIY, in videos lasting thirty to sixty seconds.[21] Humour, gaming, technology, music, sports, health and travel all came ahead of politics, in a graph that looked suspiciously close to a power law.

A lot of pain might have been avoided had people taken notice of The Onion, a satirical website, which had already summed up how people felt about video 'articles' back in May 2014 in a piece titled 'Christ, Article A Video'.[22] It took the form, naturally, of a one-minute video, and summed up how annoying people found news videos that hid behind an innocent-looking link.

'I clicked on the story thinking I'd be in and out of there in thirty seconds, but then that little buffering thing popped up, and I was, like, "Here we fucking go,"' says one person in a spoof interview. Another says, 'Look, I don't want to watch a report, or interview, or panel discussion, or whatever the hell this is. I just want to skim an article real quick and get on with my day. Put a little camera icon on [the link] or something. Don't bait and switch me.' (By capturing fundamental truths, the videos on The Onion have retained their value long after the many pivot-to-video sites' work has vanished.)

Meanwhile, news sites counted the cost of listening to Mark Zuckerberg. The *Atlantic* calculated that from 2016 to 2018 national media companies cut 350 jobs, in some cases firing writers to hire video makers, and then firing the video makers.[23] 'The very labour-intensive nature of video meant the newsroom [at the news site Fusion] had fewer people out reporting and more people making text slide in and out of the frame,' observed the *Atlantic*'s Alexis Madrigal and Robinson Meyer.

Video turned out not to be the saviour of publishing; and nor, by extension, was Facebook. By 2018, the number of staff in American newspapers had nearly halved to just under 38,000, while print adverts were worth $9.3 billion, out of a total advertising spend of $14.3 billion, the rest having shifted to the papers' web pages. In all, 1,800 papers out of 8,900 had shut between 2004 and 2018.[24] Total daily circulation had halved from 56 million per day to 28 million. A total of $40 billion in advertising revenue and tens of thousands of jobs had vanished.

The final act of the debacle came in January 2018. Facebook announced exciting changes for the News Feed: there would be more posts from friends and family members, more emphasis on 'meaningful interactions between people', but also 'less public content, including videos and other posts from publishers of businesses'.[25] The 'pivot to video' had gone through the revolving door and back out again. That came all too late for the sites, such as Mic, Vice and Mashable, that had poured resources into longer videos in the mistaken – or misled – belief that that was where 'user engagement' lay. There was bitterness aplenty among media organisations, and especially those who had seen friends and colleagues fired, or been fired themselves, in the pursuit of a mirage.

Part of the calamitous 'pivot to video' came just as companies in the US were scrambling to cover the presidential election – yielding the space to other news organisations that knew that video wasn't where it was at, or smaller ones that couldn't afford it.

The distraction for news sites of trying to make the doomed video strategy work also opened up a fertile space for 'fake news' – made-up or wildly distorted and exaggerated stories and headlines specifically intended to get people to click on them so they could be bombarded with adverts. Fake-news creators stormed Facebook from the end of 2015 and rode the 2016 US electoral cycle.

In one respect, they were aiming for fertile ground, because social networks are excellent places to spread misinformation. In mid-2018, a study by researchers at MIT looked at how truth fared against falsehood in the world where you could effortlessly pass on information to your network.[26] On Twitter, the retweet – the original form of virality – turned out to be the villain of the piece.

The study looked at how rumours spread on Twitter between its beginnings in 2006 and 2017. That involved examining about 126,000 'rumour cascades' (tweets that pushed a rumour, whether true or not, to the network) from roughly three million accounts. The more accounts retweeted the original information, the bigger the 'cascade'. (If two different accounts separately tweeted the same information – such as a tsunami warning – those counted as separate cascades.) The 'true' and 'false' labels attached to each cascade was determined from the consensus of a number of fact-checking sites. The researchers investigated seven categories, in growing order of the number of cascades: natural disasters, entertainment, science and technology, terrorism and war, business, urban legends and politics. The latter two made up more than half of the total.

The findings were unequivocal, and remarkable. The saying often attributed (wrongly) to Winston Churchill that 'a lie can be halfway around the world before the truth has got its boots on' was empirically demonstrated. False stories spread further, faster and deeper into networks than true ones. 'Whereas the truth rarely diffused to more than 1,000 people, the top 1 percent of false-news cascades routinely diffused to between 1,000 and 100,000 people,' they noted. False news reached more people, and so was more likely to be retweeted; false political news particularly so, moving through the network three times faster than any other sort of misinformation. The researchers were even able to put a number to how much more quickly a lie could spread than the truth: reaching 1,500 people took one-sixth as long for a falsehood than the truth. This was true even after removing 'bots' from the sample, as they turned out to be just as assiduous about passing on truth as fiction.

As might now feel predictable, the most common emotion expressed in response to the false or true tweets was disgust – a particular form of outrage that was stronger for the false news, and thus made it more likely to go viral.

By 2018, more than a quarter of all tweets were retweets, and analysis suggested that the proportion was steadily rising as people found it easier to pass something on with a single gesture than to write a piece of original content.[27] The implications for accuracy weren't encouraging.

Just four days after the shock result of the 2016 US presidential election, Mark Zuckerberg was insistent that fake news on his site could not have made a difference to how people voted. 'The idea that fake news on Facebook, which is a very small amount of the content, influenced the election in any way, I think is a pretty crazy idea,' said Zuckerberg, then thirty-two, at the Techonomy conference in San Francisco.[28] 'There is a profound lack of empathy in asserting that the only reason someone could have voted the way they did is because they saw fake news.'

His refutation of the idea that 'fake-news' content made a difference was 'why would you think that there would be fake news on one side but not on the other?' To many onlookers, this suggested he hadn't been paying much attention in the past year. 'Facebook stumbled into the news business without systems, editorial frameworks and editorial guidelines,' Claire Wardle, then of the Tow Center for Digital Journalism, told the *Guardian* in an article published the day before Zuckerberg spoke.[29] 'And now it's trying to course-correct.'

An analysis by Craig Silverman of Buzzfeed News of the fifty most-shared fake-news stories of 2016 on Facebook – articles that were '100 percent false and that originated on fake-news websites', rather than simply being partisan misrepresentations – found that twenty-two were about American politics.[30] For an election year, that's not surprising. What was telling was that twelve of them were either pro-Trump or anti-Clinton, and had racked up nearly 7.5 million views, while the six pro-Clinton ones mustered just 1.8 million views, and the four neutral ones (example: 'Obama passed law for grandparents to get all their grandchildren every weekend') collected 1.1 million. Zuckerberg had been wrong: the fake news was indeed more on one side than the other. It also might have been simply giving people what they wanted, but there was a feedback process at work. As we've seen before, the shock factor was surely important: the fake story that Pope Francis had endorsed Trump for president got nearly a million views, while a version of the same story suggesting he had endorsed Clinton for the same job got one-fifth as many – a sign that people weren't as surprised by that idea, imaginary though it might be.

The purveyors of fake news had simple motives for getting people to click through to their stories: they would get a slice of the revenues from the adverts there. Just like supermarkets tabloids such as the *Weekly World News* (famous for outlandish front pages such as 'World War II Bomber Found On Moon'), creating fake news was the front-and-centre purpose of the enterprise. The supermarkets that sold the paper were indifferent; as long as it wasn't illegal, they didn't mind. Facebook took a similar view: people were free to believe this stuff, just so long as they kept coming back to the site. Politics and influence was somebody else's problem.

Yet there is a contradiction here that neither Facebook, nor Zuckerberg individually, acknowledged. There was scientific proof that Facebook could influence people to go out and vote. And there was also a scientific demonstration that the content of the News Feed could manipulate people's emotions. Why shouldn't fake news showing up in the News Feed and outraging people prompt them to get out and vote? It's social warming: the effect doesn't have to be big to make a difference.

Furthermore, an online survey carried out by Ipsos Public Affairs just after the election disproved Zuckerberg's implication that fake news – particularly about politics – didn't catch people's eyes.[31] Presented with eleven headlines from the election, some true and some false, on average about a quarter said they had already seen them. But when asked to judge their truth or falsity, about three-quarters thought the fake-news stories were true; more if they also cited Facebook as a major source of their news. There was also a partisan split, with a statistically relevant difference: 84 percent of the time, Republican voters said the fake headlines were accurate, against 71 percent for Democrats.

Obviously, fake news plays on the credulity of the reader. And news organisations have pushed nonsense on their readers for years. But fake news was a natural side effect of the design of Facebook, Twitter, Instagram and other networks. A thumbnail picture with an arresting headline and a plausible name or URL ('Denver Guardian', 'Christian Times', 'politicops.com') was as convincing as any from a legitimate site. Publishers could, and still can, edit the headline text attached to a link to amplify its effects, or create multiple versions to see which one resonates better. (Users were blocked from doing this in mid-2017.) Given that 44 percent of people got their news from Facebook, the problem for everyone – but also the opportunity for the fake-news vendors – was obvious.[32]

In the conversation at Techonomy where he dismissed the problem of fake news, Zuckerberg said: 'My goal, and what I care about, is giving people the power to share and giving every person a voice so we can make the world more open and connected. That's the mission.' Yet it seems a bit of a stretch to call fake news, whose authors don't care if it misleads, part of 'giving people the power to share', unless you're indifferent to the effects of misleading them.

So what are the effects on people of fake news? Generally, you'd expect that you'd see a lowering of trust in institutions and information sources that don't agree with what the person wants to hear.

We can measure that, at least in the US. The polling organisation Gallup has since 1979 run an annual survey asking people about their levels of trust in organisations such as the church, Congress, big business, unions, newspapers, the presidency, banks, TV news and other institutions.[33] Occasionally people are asked about trust in 'news on the internet' (three times so far, in 1999, 2014 and 2017). The survey asks about the amount of trust people have in the various institutions – a 'great deal', 'quite a lot', 'some', 'very little' or 'none'.

The picture isn't encouraging. If you add the enthusiastic responses ('great deal' plus 'quite a lot') and subtract the most negative ('very little' or 'none'), you get a broad measure of trust, or distrust. So if a total of 60 percent of people have a great deal or quite a lot of trust, and 15 percent say 'very little', the trust figure is 45. For most institutions, that number has been drifting down for decades. For all but the military and small business (both of which are cumulatively above 60), sentiment has become markedly worse since about 2003. The US Congress first moved into negative territory in 1991, and after a positive span from 1997 to mid-2005 has careened downhill to a −37 mark in June 2019; it is the institution with the worst mark.

News, in general, hasn't fared much better. Both newspapers and TV news saw sudden collapses in positive sentiment in 2003 – around the time more than 60 percent of Americans had internet access. By 2006, sentiment had turned negative, but the sharpest decline began after 2011, by which point more than a third of adults had a smartphone and social media use passed 50 percent.[34] By mid-2019, TV news scored −27, and newspapers −11.

Just a week after his confident Techonomy appearance, Zuckerberg changed his stance. As Facebook absorbed the fallout of its role in the 2016 election, and the failure of its own moderators to keep abreast of the avalanche of hoaxes and fakes, it decided to turn to outside organisations – the growing number of 'fact-checking' sites that had grown over the years as the amount of misinformation and invented stories on the net had blossomed. In a Facebook post a fortnight after the election, Zuckerberg insisted that 'We believe in giving people a voice, which means erring on the side of letting people share what they want whenever possible. We need to be careful not to discourage sharing of opinions or to mistakenly restrict accurate content. We do not want to be arbiters of truth ourselves, but instead rely on our community and trusted third parties.'[35]

It was a neatly executed body swerve towards making the site more trustworthy, while never actually whispering the word. Facebook doesn't want to be an arbiter of truth because that is completely unimportant to its mission. Truth, lies, misinformation: all are grist to the mill of showing more adverts and keeping people on the site longer.

So what would happen to content that was judged inaccurate, or misinformation? 'Similar to clickbait, spam and scams, we penalise this content in News Feed so it's much less likely to spread.'

Consider that sentence: Zuckerberg is saying that spam and scams *aren't removed* from Facebook, because they're not offensive – unlike, say, breastfeeding mothers' nipples. Content intended to rip people off and deceive them is acceptable; it just won't be given a boost by the algorithm that oversees everything. And now misinformation was being put into the same box: permissible, tolerated.

In December 2016, Adam Mosseri, then in charge of Facebook's News Feed, announced the fruits of the attempts to control intentional lying (or perhaps 'lying for profit'). He repeated Zuckerberg's phrase about believing in 'giving people a voice' and not being arbiters of truth. Then he announced who would be: in future users would be able to report content as 'fake news', an addition to the previous options of 'annoying/not interesting', 'shouldn't be on Facebook' and 'spam'.[36]

Mosseri explained that 'We'll use the reports [of fake news] from our community, along with other signals, to send stories to these [fact-checking] organisations.' Those would then decide if the stories were untrue; if so, Facebook would flag them as 'disputed', and downrank them in the News Feed algorithm.

It sounded Batman-esque: the Fake News signal would go up in the night sky, and fact-checkers would leap to their terminals and validate a story, or wrestle it to the ground and slap on handcuffs marked FALSE.

The reality was more mundane, as Brooke Binkowski explained to me. At the time of Mosseri's announcement, she was editor-in-chief of Snopes, which is one of the oldest fact-checking sites, dating back to 1994, and which was one of the first in the new partnership with Facebook. The first fact-checks began appearing on stories in March 2017.[37]

'They would provide us with this list, for each of us personally, that was hooked into our personal Facebook accounts. They showed you a URL to a story, and we would mark whether we had found it true or false, and put a link to our own debunking of it. Then move on to the next one.' There was no quota, though there were usually at least 250 stories to be refereed, divided into ten pages of twenty-five. The list of stories was unending, since you'd never run out of people being wrong on the internet. 'The cup was continually full,' Binkowski recalls.

On Facebook, fake items would be given a small tag, and anyone who tried to share them would see a small warning about its disputed veracity – though they could still share it. After a while Binkowski began wondering how the stories were chosen, suspecting that they were overwhelmingly being flagged by an algorithm rather than humans. Not only that, but they had a strange uniformity.

'It was weird. Often what they were showing us was the same story but slightly reworded, or the same story but from a different site. I kept asking [Facebook] what the criteria were [for stories to appear for fact-checking]. We were getting a lot of stories about stuff that was considered "left-wing", so there was anti-vaxxer stuff [opposing vaccination], stuff that was anti-guns, and then a lot of stuff that, even if people fell for it, really wasn't corrosive [in a political sense]. It was just muddying the waters.'

Facebook wasn't forthcoming with her about the criteria by which stories were chosen, and seemed powerless to prevent variations on the same story being presented again and again.

The problem, Binkowski came to realise, was that this was an asymmetric war. Within the parameters Facebook had set, there was no way to stem the flood of misinformation, disinformation and untruths. The penalties for posting fake news were minimal, because the content wasn't deleted from Facebook, only 'downranked'. By contrast, the potential rewards from spreading it were substantial, and generating new versions of the same story (which looked to Facebook's algorithms like entirely new content) by writing a new headline and tweaking a few words and a picture was cheap.

Facebook brought on four fact-checking organisations as partners. They stood as much chance as Canute. Looking back on the experience later in an article for Buzzfeed News, Binkowski described it as 'the world's most doomed game of whack-a-mole, or like battling the Hydra of Greek myth. Every time we cut off a virtual head, two more would grow in its place.'[38]

By September 2017, Zuckerberg had definitely changed his mind, at least about fake news's impact. The company admitted that Russian operatives had spent $100,000 on adverts aiming to widen the existing social and political splits in the US.[39] Donald Trump immediately took to Twitter to claim that that was irrelevant, because 'Facebook was always anti-Trump.' Perhaps stung at the idea that his creation hadn't had any effect, Zuckerberg responded that he had personally been wrong to dismiss the potential impact of what people saw on Facebook, including fake news.[40] 'This is too important an issue to be dismissive [about],' he wrote in a post on the site, apparently forgetting his dismissive comments less than a year earlier. 'But the data we have has always shown that our broader impact – from giving people a voice, to enabling candidates to communicate directly, to helping millions of people vote – played a far bigger role in this election.'[41] He didn't offer to share the data.

Again, though, the point about warming, whether social or global, is that it doesn't have to be big to have an effect. And Zuckerberg was overlooking the advantages that the creators of 'fake news' have over purveyors of factual news: it's easier to generate fake news than the real stuff, because you're untrammelled by reality and you can thus make it both unique and eye-catching, both of which promote its spread.

At the end of 2017, Buzzfeed News once again examined the scale of the problem of fake news on Facebook.[42] The fifty most viral fake stories of 2017 had more 'engagement' (shares, Likes and comments) than the top fifty for 2016, when the world had woken up to the breadth and depth of the problem. The problem was getting worse, not better: the Buzzfeed data showed that the top fifty had nearly 10 percent more engagement than the previous year. With the election over, politics had eased as a category, down to eleven in the top fifty; now weird 'crime' stories dominated.

Facebook was faintly dismissive of Buzzfeed's findings, implying in a statement that things would have been much different if its fact-checking systems had been in place for the whole year.

That, however, ignored the reality. Fact-checking could take up to three days, and the vast majority of clicks would occur before then. That's the nature of virality: surf the wave, and then wait to catch the next one. Buzzfeed's analysis found that for every click on a fact-check, a hoax got 200. Truth was losing out to lies. 'We know we miss many [fake-news stories]', admitted Jason White, then Facebook's manager of news partnerships, in an email to the fact-checking companies.[43] 'We must also take additional steps to meet this challenge.'

For a fake-news publisher, the only criterion was viral success to drive viewers, and thus advertising eyeballs, to their site. The fact-checkers and the fake-news sites were working to different rules, at different speeds; yet Facebook was acting as a gearbox yoking them together, and claiming the benefits of both. Facebook benefited from the engagement that the fake-news sites produced, as bamboozled users clicked on stories with headlines such as 'Babysitter Transported To Hospital After Inserting A Baby Into Her Vagina' and 'Morgue Employee Cremated By Mistake While Taking A Nap', while also burnishing its reputation by pointing to the fact-checkers' efforts, and getting to show adverts to users.

In fact, the fact-checkers' efforts probably had the opposite effect from what was desired. In September 2017, a team at Yale, MIT and the University of Regina had found that while putting a 'disputed' tag on stories made people only marginally less likely to believe them, the lack of a tag on other stories would make them assume these stories had been checked and were correct – what the researchers called 'implied truth'.[44] Given the average three-day lag in picking up on fakes, that could have boosted rather than quelled the fake-news boom.

Complicating the matter of fact-checking even further was that fake-news sites discovered a Get-Out-Of-Jail card if they were accused of being misleading. In March 2018, a site called the Babylon Bee ran a story titled 'CNN Purchases Industrial-Sized Washing Machine To Spin News Before Publication', which was duly fact-checked by David Mikkelson, Snopes's co-founder and owner. He labelled it False. Facebook warned the Babylon Bee that publishing more 'disputed information' would see it downgraded in the News Feed, and it wouldn't get any more Facebook ads.[45]

The Babylon Bee had a simple reply: of course the story is false – we publish satire. 'We're a satire site' quickly became the go-to response for sites that wanted to spread untruths, yet not be judged by the fact-checkers. Some would put descriptions on their sites saying that some of it was news, and some was satire. So which of their stories was which? Well, that depended who was asking. (The label on Mikkelson's story at Snopes about the Babylon Bee piece was changed from 'False' to 'Satire' sometime after September 2018.[46] Snopes didn't answer questions about when that was done.)

Yet for all Binkowski's frustration at the ever-full cup of fake news, fact-checking did affect those who had previously profited from using Facebook's gigantic reach to draw people to their sites.

In August 2018, the Poynter news site found a number of the players in the fake-news ecosystem complaining bitterly that their business model was being destroyed while Snopes was getting a traffic boost from the fact-checking tie-up.[47] 'Mikkelson is screwing satirists for profit and he's going to get good and sued for it,' said Christopher Blair, a middle-aged man based in Maine who ran a network of (as he described them) satire and parody sites, which were being repeatedly knocked back by Snopes. In an interview in May 2017 with PolitiFact, Blair had insisted that the only purpose of his site Last Line Of Defense was, as it said in its own description, to 'expose the extreme bigotry and hate and subsequent blind gullibility that festers in right-wing nutjobs.'[48]

But Snopes's fact-checking was destroying the ability of Blair's sites (sample story: 'Fox News star Tucker Carlson in critical condition (then died) after head on collision driving home'; the satire and/or parody may be hard to spot) to reach people on Facebook.[49] As they got more and more negative fact-checks, Facebook made sharing links to them harder and harder, until by mid-2018 attempts to do so generated a message saying, 'We believe the link you are trying to visit is malicious. For your safety, we have blocked it.'

The fake news sites did try their own form of the 'pivot to video' that so many news sites had thought might be their saviour: they did a 'pivot to graphic', generating pictures with outrageous captions embedded in them, on the basis that they would be harder for machine learning systems to decode and detect. But those too could be reported and knocked back.

The makers of the stories and graphics, particularly Blair, insisted that they weren't fake news; they were satire. 'Fake news is that horrible thing that sways elections and destroys America,' Blair told Poynter.[50] 'I have no part of that. I write fiction.'

Or, as a car driver might put it: I'm not contributing to climate change during my drive, because my journey is necessary. It's those other people who are the problem.

At the end of 2018, when the fact-checking system had been running for a full year, rather than the nine months of 2017, Buzzfeed looked again at the biggest fake-news stories.[51] The level of engagement (Likes, shares, comments) was lower than in 2017 – by 6 percent. But it was still higher than in 2016. Even if some of the players were being shuffled, the fake news problem wasn't changing at all.

Binkowski felt that the kludgey nature of the system for categorising fake news was getting in the way of what is clearly a motivating force for her: getting Facebook to be truthful and transparent. 'I thought they might be using it to adjust their algorithms but whenever I asked them, they wouldn't answer, or they'd give us some platitudes about "If you debunk something, it reduces its reach by 80 percent," and I'd say, "OK, where's the research to back this up? Show me." And they never did.' She became more and more confrontational 'until they stopped inviting me to meetings [with other fact-checking sites].' But she remained frustrated with the lack of transparency: 'It seemed like this weak weapon, like they were just giving us busywork so they could say, "We have Snopes, we have the AP [Associated Press], we have AFP [Agence France Presse]". That's really honestly what it felt like.'

She suggested, in at least one of the conference calls with Facebook, that the site could do just as well or better by recruiting more of its own users to moderate content: some people like the status of being a moderator, and by crowdsourcing to cross-check their accuracy, the best moderators would emerge naturally. 'I remember when I was on IRC [internet relay chat, one of the earliest interactive internet forums] we had moderators, and we did it for fun, and the community moderated itself. I kept telling Facebook that. I was all, like, "Hey guys, listen to the old person, listen to the forty-year-old, the Gen Xers who have been on there this whole time, about moderators."' (Her suggestion was later vindicated by an academic study in 2020 that found that fact-checkers were trying to mop up a flood with a towel: Facebook's six official fact-checking partners were deploying twenty-six full-time staff who could check, on average, about two hundred pieces of content per month, or ten per working day. The study found that a politically balanced panel of Facebook users could do the job just as well – and much faster.)[52]

Facebook didn't take up Binkowski's suggestion. Instead, the company continued to insist that the fact-checking arrangement was good for all concerned, even when both Snopes and the Associated Press withdrew from it in February 2019, making noises about resources; the unspoken complaint seemed to be that Facebook wasn't paying enough.[53] For Snopes, vice-president Vinny Green told Poynter that the process was still manual and time-consuming. He suggested that Facebook was, in effect, wasting fact-checkers' time, and that fake-news sites should simply be reported by different methods so Facebook could get them off the site.[54] 'At some point,' Green told Poynter, 'we need to put our foot down and say, "No. You need to build an API"' – a data feed that could be used to build better tools for checking content. That, suggested Green, could then be used among all the fact-checking sites to benefit all of the web.

Splitting from Facebook wasn't cheap: in 2018 one-third of Snopes's total $1.22 million budget came from its fact-checking payments.[55] Without that, and a 'shareholder infusion' of $500,000, the site might have gone under, or drastically cut staff. (In 2019, the budget rose to $1.28 million, of which $1.16 million came from crowdfunding and reader contributions.)

The decision to break away from Facebook illustrated the problem that the social network was keen to avoid shouldering itself: while machine learning could nominate potentially bad content, humans had to make the final decision about every article.

Facebook was oddly happy to rely on its systems to make work for humans, but reluctant to let those same systems block content. This was the reverse of its approach to the creation of millions of fake accounts every day, which were overwhelmingly spotted and blocked by another set of machine learning systems without human intervention. That is a matter of priorities: the advertisers who fund Facebook won't pay for adverts if they think they're being shown to non-existent people, so Facebook has a direct financial interest in ensuring that every account is tied to a real person. By contrast, if real people choose to read and share utterly untrue content, that isn't a problem for Facebook or its advertisers in the slightest.

Like Google, though, the algorithms in charge of what you see on Facebook and Twitter aren't attuned to showing you what's *true*. They're trying to pick what you and people like you find *engaging*: the stuff that you spend time looking at (whether in delight, disgust or disbelief), the stuff that you then pass on. 'Sometimes you open Twitter and there's not very much of interest,' says DiResta. 'But sometimes you open it and there's a tweet that's thirteen hours old, but because it's become a huge morass, a huge battleground, that's what is up at the top of your feed. That's because in a ranked-choice algorithm, that's the most interesting thing on the platform.'

The role of the algorithms is important, DiResta says. 'If you spend most of your time on the internet in a mom group, then what you're seeing is what the platform algorithms have decided you should see. That's usually just high-engagement content, where you start to see that things that are sensational, that are emotionally resonant, are going to perform better. The problem is, we haven't yet figured out a way to say, "Should it be showing us other things as well? Shouldn't it be highlighting more authoritative sources somewhere in this process?" That's part of the problem: people have freely chosen to find themselves in these groups, but we've never before had a system where that [algorithm dependency] is the dominant environment, and reputable information is never even getting in there.'

For Binkowski, the experience with Facebook's fact-checking hasn't made her stop wanting to fact-check the internet. After a contentious departure from Snopes in July 2018 – she says that Mikkelson gave her a number of conflicting reasons why, ending with the suggestion that her job had ceased to exist as part of a restructuring – she joined another long-running fact-checking site, Truth Or Fiction?, bringing along other ex-Snopes staffers. 'I can't even get on Facebook any more, it's so depressing,' she told me. 'It just all angry right-wingers yelling at each other and passing each other these horrible false stories and then railing at the press.'

She still sees Facebook's externalising of its fact-checking programme as a means of avoiding responsibility at a cut price. 'It's cheaper. They don't have to pay in-house people, and it's easier to discredit outside entities and, I believe, easier to control them or fire them or get rid of them if they're not controllable. At first, they were offering to pay all the fact-checking sites a hundred thousand dollars a year. If you've got twenty fact-checking organisations and you're paying them a hundred thousand dollars each, that's still way cheaper than a crisis PR firm. It's still cheaper than hiring moderators and journalists in-house, than giving them health insurance. It was just optics. I think that's what they did it: because it took the heat off them.'

To some extent, the public's willingness to accept fake-news sites as potentially real had been enabled by news sites themselves as they struggled to adapt to the new world where the social networks didn't just provide a few extra readers, but actually determined their fate. Publishers sought new sources of advertising revenue that weren't subject to the capricious whims of Facebook, nor the low-paying Google AdSense. Ready and waiting for them were the 'content marketing networks', or CMNs – companies such as Taboola, Revcontent, Outbrain and Adblade.

The tie-up turned out to be lucrative, but damaging in the longer term. The CMNs paid handsomely, by internet advertising standards, and tied up deals with reputable news organisations, including technology sites such as Fast Company and Wired, and longstanding (but struggling) brands like *TIME*. In return, they plastered the space below legitimate news stories with boxes titled 'Promoted Stories' or 'Around the Web' and emblazoned with look-again photos and clickbait headlines. And what headlines. 'One Weird Trick To Reduce Belly Fat', 'Tom Selleck Makes Brave Statement About his Personal Life', 'Meet the Women Making Rape Jokes That Are Actually Funny'. Sometimes the 'news' sites were paid advertorials; some were actually schemes to collect emails via 'registration' so that people would be bombarded with spam.

By October 2016, an analysis by ChangeAdvertising, a non-profit, found that forty-one of the fifty biggest news sites were using such 'content ads'.[56] A quarter of the ads were categorised as 'clickbait' ('1 Weird Trick That Forces Your Eyes Into Perfect 20/20 Vision In Just 7 Days'), and nearly all were from sites whose ownership was murky at best.

But who cared, if people were making money? Outbrain claimed to be generating up to 30 percent of revenues for some publishers, and that it saw 250 *billion* clicks per month – one in twenty of every offered link, fifty times the industry average for banner ads. Others could count on payments of more than $1 million per year, as long as they let Taboola have its way. The best thing? A few lines of code on each web page would pull in the content from the CMN's servers whenever a page was loaded. For the publisher, the deal was pure profit. The only challenge was meeting page view targets to show the ads, which could mean including them all over the site.

But this was an example of measuring success only by what you can directly measure, and ignoring the importance of sentiment and reputation. Readers of upmarket publications saw the CMN ads, and they weren't impressed.

Damon De Ionno, chief executive of the British consultancy Revealing Reality, says the ads were a self-inflicted wound. 'The idea of the media brand which takes responsibility for the provenance of what it has published has been undermined,' De Ionno told me. A physical newspaper would take a lot of care about the reputation of advertisers in its paper; those known to be scammy would be refused. When they moved online, though, the money on offer was too good to refuse.

By the end of 2016, publishers had already begun to rethink their position, with some of the larger ones, such as the *New Yorker* and *Slate*, abandoning them.[57] Publishers began to recognise the damage that was being done. In 2017, an investigation by *Wired* demonstrated that CMN's original intent, of helping small publishers reach big audiences on respected sites, had been superseded.[58] Instead, they had been used to push fake news, which inevitably in 2016 had meant that the usual fare ('Kate Middleton Drops James, Proving Prince William is a Lucky Man') was replaced by fake news about Hillary Clinton and the US presidential campaign ('I Supported Hillary Clinton Until I Saw This'). 'We hate this, but we can't walk away from the money,' one told Digiday.[59]

But the damage had been done, De Ionno says. Media content had become a sort of visual porridge, where any but the biggest brands struggled to stand out or gain trust; but equally, any fly-by-night could pretend to be authoritative. 'The presence of any sort of logo on a news site often, for some people, counts as "that's a thing".

To them, that could be as good as the BBC. And on social media it's easier to browse without engaging fully, and just do more skimming.' He contrasts it with the idealised world that existed before news was atomised and organised into echo chambers: 'It was prepackaged. Someone had thought that this story and that story together were useful and balanced, and that did have some sort of quality control. Now we have put the power in the public's hands, and that human laziness kicks in. It's like the way that it's easier to go and get fast food rather than cooking for ourselves.'

Any expectation that people might improve their news diet after the experience of the past few years was dashed by a study in October 2020 that found engagement on Facebook with 'deceptive' news sites was twice as high in the run-up to the US election than in the same period in 2016.[60] For some specific sites, it had tripled. The concentration was notable: ten sites, including Breitbart News, The Blaze, Daily Wire, DJHJ Media and The Federalist, accounted for nearly two-thirds of interactions. Almost all were right-wing, catering to a hectoring vein of American political discourse. 'There is a need to fill the vacuum left by the decimation of journalism with a new PBS [America's independent publicly funded broadcaster, resembling the BBC in the UK], funded by a fee on online ad revenue,' the researchers suggested. Facebook responded that analysing Likes, shares and comments was misleading, because it didn't represent what 'most people' see on the site. As so often, though, that is dissembling (and Facebook offered no other data to back up its argument). The point isn't what most people see there, but what is impactful and foments narratives. A Facebook without those sources would immediately be a calmer, less divisive place. But that is self-evidently not what the company wants.

Meanwhile, having a Twitter account had become a journalist's most indispensable tool. Striking the right balance between looking for stories on Twitter – where contacts and discussion could be happening – and slogging through phone calls and emails and fruitless research could be difficult: Twitter was temptingly always there, and there always seemed to be a story of some sort breaking, even if it wasn't a plane ditching in a river. Yet the way in which Twitter effectively allowed journalists to wander off the homestead prescribed for them by their employers created new headaches for both writers and publishers. If a non-political journalist weighed in with personal, political opinions, was that the publication's viewpoint? Sometimes tweets could cause legal peril: at the *Guardian*, I once retweeted a Wikileaks tweet about a company called Trafigura – blissfully unaware that doing so would break an injunction that applied to everyone at the *Guardian*. The injunction's existence was itself a secret to all but a few executives. The lawyers had reasoned that if anyone on the paper wrote a story about the company, the executives could intervene before it was published. They hadn't considered social media.

In the US, the bigger risk from Twitter seemed to be how much personality a reporter could, or should, inject into their tweets. There, journalism has long attempted to provide what Jay Rosen, professor of journalism at New York University, calls 'the view from nowhere' – a form of objectivity that strives beyond all else never to pass an opinion about the people and events being written about. Thus a remark about 'shithole countries' by the president is not described as 'racist', but 'racially charged'; climate change is not a scientific fact, but something for two sides to argue about, regardless of accuracy. It's not just impartiality; it's an approach that holds adjectives at arm's length in tongs. Twitter changed that; journalists could court popularity or unpopularity; abuse was common. By 2017, news sites were rejecting people for jobs on the basis of 'inappropriate, partisan or puerile' tweets they had written in the past.[61] You could also be fired.

The 'view from nowhere' collided dramatically with the encouragement of reporters to be on Twitter in order to boost their organisation's visibility in 2020. One American paper banned a reporter from reporting on the Black Lives Matter protests during spring 2020 because she had jokingly tweeted a picture comparing the aftermath of a concert by Kenny Chesney, a white country singer, to that of looting.[62] At the *New York Times*, several journalists concerned by a comment article written by a right-wing senator, Tom Cotton – which called hyperbolically for the use of troops to stop 'rioters' in 'many American cities' – complained on Twitter that 'Running this [article] puts Black @nytimes staffers in danger.' One of them, Nikole Hannah-Jones, tweeted that she was 'deeply ashamed that we ran this'.

Newsroom revolts are unusual, but Twitter gave the row added velocity and publicity. Within days, there had been resignations at the top of the paper's opinion section, in which the controversial piece appeared – and there was a huge cottage industry of people writing articles about how the *New York Times* had been right, wrong and everything in between to run the article and accept the resignations. No one, however, questioned whether the journalists should have been allowed to tweet their opinions; that had become part of the media landscape. Give journalists a blank space to type into and they'll often find the temptation unendurable. Twitter provided the spark that lit the flames of dissent.

Attempts to hold the line looked like dams under stress. At the *Washington Post*, the other major American national newspaper, an internal review in April 2020 tried to square the circle of the conflicting demands now being made of staff on the 'national desk', the section that includes news reporters.[63]

The proximal cause of the review was the suspension in January that year of Felicia Sonmez, a news reporter. Just hours after the death of the basketball star Kobe Bryant and his daughter in a helicopter crash, she had tweeted a link to an article from 2016 in the Daily Beast, an online publication, that described details from a 2003 accusation of sexual assault against Bryant. Sonmez hadn't been assigned to the Bryant story; she wasn't writing about it, wasn't a sports reporter. Having previously experienced sexual assault, she later said she felt it was 'jarring' that there was no mention in the early obituaries about Bryant of the case.[64] She suggested that her tweets didn't undermine her colleagues writing about Bryant; instead, 'it demonstrates to [assault] survivors that we see them and hear them, and they are not ignored.'

Twitter's inhabitants didn't see it that way, and Sonmez was targeted for abuse by thousands of accounts, including threats that mentioned her home address. But rather than being supported, she was told by her managers that she had gone beyond her 'coverage area' and 'undermined' her colleagues' work, and suspended. That triggered another outcry, which also played out on Twitter, through the publication of a joint letter by the journalists' union in the paper. Sonmez was reinstated, but the paper's social media policy was in ribbons.

The *Post*'s internal review spoke to more than fifty of its reporters. The review confirmed that, for reporters, being on Twitter was a demonstration of the paper's digital-first attitude, a necessary reporting tool, a method of sharing other stories from the paper, and a way to add context to their published work. Dropping back would mean losing out: 'Because I'm not as active on Twitter anymore, I'm not booked on TV as much,' one reporter told the review. Some saw Twitter as a sort of escape hatch to a better future. 'The very top [of the paper] is already very crowded, and if you want to be noticed, maybe it's better to do it somewhere else – like social media,' one said.

Another pointed out the obvious. 'I feel like there's an inherent contradiction in social media. You're asked to have lots of followers and to not be opinionated. But the way you get lots of followers is to lift up your skirt a little.'

Another said: 'If I'm deathly afraid of driving, I can opt not to drive. I can take the bus. I feel like I don't have the option to opt out of tweeting.'

Yet another said that the real-time, publicly visible nature of news now meant that 'the reporting process is just inherently messier now than it was in the Olden Days.' The solution? There wasn't one. 'We should acknowledge it and get our readers used to it,' the writer said.

The end of 2019 provided a perfect illustration of how social warming, through the distorting lenses of Facebook and Twitter, had affected journalism. On Sunday 8 December, with just four days to go before voting in the UK's 2019 general election, the dull news agenda lit up. A photo of a four-year-old boy with suspected pneumonia who had had to lie on the floor of the emergency department at Leeds General Infirmary hospital began spreading on social media. He was on the floor because there were no beds available; he had to lie on a coat because he couldn't get comfortable on a chair. The photo, showing a transparent line to an oxygen mask snaking down beside him, accompanied a story in the *Yorkshire Evening Post* that also included an apology from the chief executive of the hospital.[65] The child's mother said the staff had been 'all as helpful as they could be' but the NHS was in crisis, as evidenced by her child spending twice the nationally mandated waiting time in the A & E department, and only getting a bed on a ward more than thirteen hours after arriving.

On Monday morning, the story and the picture were printed by the *Daily Mirror*, a national left-wing daily paper. Yet despite its unequivocal facts – the photo, the multiple apologies from hospital managers – the story was quickly disputed, but not by the hospital or any authoritative or political sources. Denials began appearing on Facebook and Twitter, tending to take the same form. The original was: 'Very interesting. A good friend of mine is a senior nursing sister at Leeds Hospital. The boy shown on the floor by the media was in fact put there by his mother who then took photos on her mobile phone and uploaded it to media outlets before he climbed back onto his trolley.' Inevitably, the Facebook version of the post called the image 'fake news'. Other versions of the post disputed the function of the oxygen line, and whether someone with suspected pneumonia would be given one.

The posts went viral on Facebook, particularly among Conservative and Brexit supporters, as it fitted everything they wanted to hear: that stories of an NHS in crisis were exaggerated, that the left-wing media was making stories up, that mothers who were describing an NHS in crisis were liars in close contact with the media. But the 'debunking' was false. There's no 'Leeds Hospital' (the hospital was 'Leeds General Infirmary', and medical staff tend to be picky about such naming), and there's no such title as 'senior nursing sister'. But people didn't bother to check. Five Conservative candidates spread the 'debunking' on Facebook and Twitter, suggesting the photo was faked. The woman on whose Facebook account the original 'good friend of mine' post appeared later told the *Guardian* that her account had been hacked: 'I am not a nurse and I certainly don't know anyone in Leeds,' she insisted.[66] She had received death threats because of the post, she said.

But the plight of the boy, and the hospital's struggles, had pushed the NHS up the news agenda. Late on the Monday afternoon, the health minister and Conservative candidate Matt Hancock visited the hospital. It was a PR effort that could have no real impact on staffing or funding, given that winter was beginning to bite. But as Hancock left the hospital in the gathering darkness with his staff and walked to his car, something happened.

Military strategists talk about the 'fog of war' – the impossibility of knowing quite what is happening once hostilities begin. That evening, the fog became indistinguishable from a red mist.

At about 4.30p.m., a number of high-profile political journalists suddenly tweeted what seemed like a dramatic story: a Labour 'flash mob' had scrambled to the hospital and 'it turned nasty when they arrived – one of them punched Hancock's adviser', tweeted Laura Kuenssberg, the BBC's political editor. Robert Peston, political editor for the ITV network, tweeted that the adviser had been 'whacked in the face by a protester'. Tom Newton Dunn, then political editor for the *Sun* newspaper, went even further: 'Today is getting nasty. Matt Hancock's special adviser has been allegedly punched by a flash mob of Labour supporting activists outside Leeds hospital', he tweeted.[67]

Suddenly, the election campaign seemed to have hit ignition point: the two sides were actually physically fighting, and Labour was the instigator – a story that fitted well with the line that the Tories wanted to take, which was that Labour posed a risk to the country. Not only that; it was even whispered that Labour had paid for the taxis that had taken the flash mob to the hospital.

All three journalists said they had heard the claim of the punch from two sources (a standard criterion for journalist verification), which made the tight grouping of their tweets surprising: finding two independent people who were on the scene of an event is usually difficult. The reality, it quickly emerged, was that both of the sources were Tories, though none was ever named.

When – inevitably – video of the incident emerged about an hour later, it turned out that the claims were wildly overstated. There had been no flash mob. There was a single angry activist, wheeling his bicycle, who had harangued the minister and followed him to his car, and then continued gesticulating as the car had sped off, throwing his arm out. As the activist looked back while shouting, the Tory adviser had walked past – and into his arm. The glancing, unintended blow had surprised both of them.

But the video took time to emerge, whereas tweets can go viral. Marc Owen Jones, an assistant professor in Middle East Studies and Digital Humanities at Hamad Bin Khalifa University, analysed the spread of tweets, particularly from 'influential, verified accounts'.[68] Newton Dunn was the first to tweet the claim, and had more than a thousand retweets. The story became a fire that was only slowly put out as the video spread, around two and a half hours later, and those who had tweeted about it in the first place recanted – and, in most cases, deleted their original inflammatory tweets.

To many journalists away from the action, the political editors looked like hopeless dupes. Their Tory sources had spun them a line, and they had breathlessly tweeted it without checking accuracy by, for example, getting the input of a bystander. Didn't facts matter? Why was it so important for the BBC or ITV to tweet the claims, rather than investigating them properly?

Some saw the incident as raising fundamental questions about disinformation and manipulation, for which they blamed Dominic Cummings, who had driven the Vote Leave focus on Facebook adverts, and was now coordinating the Conservative campaign. David Yelland, a former editor of the *Sun*, tweeted that 'Dominic Cummings skilfully manipulated the media yesterday, creating a fake news brawl around Matt Hancock and then attacking the BBC: all to take attention away from a child on the floor of a hospital. It worked and won the day. But why are current editors allowing this?'[69]

Whether Cummings was one of the 'sources' isn't known. But even if he was, he couldn't force Peston, Kuenssberg or Newton Dunn to tweet anything. What, and how quickly, they chose to write was entirely up to them. Both clearly felt impelled to push out what everyone else perceived as 'news' – stuff I want to pass on – as soon as they got it.

This was a perfect example of social warming in full flow. In the past, neither Peston nor Kuenssberg nor Newton Dunn could have spread the initial incorrect message. As reporters on TV networks and on newspapers, they would have had to wait either for camera footage, or for the printing of an edition (or, more recently, access to a computer, to write a web update). That would have meant a delay during which they could have checked the accuracy of the story they were being told, and discovered the less dramatic truth. News bulletins would probably have gone untroubled.

Kuenssberg and Peston apologised, and deleted their earlier tweets. Newton Dunn left his up, adding another tweet ninety minutes after the first saying the incident was 'not an intentional punch, but a misplaced arm. Tories also now saying it might have been an accident – but the adviser felt at the time it wasn't. The most acrimonious day of the campaign so far.'[70]

The deletions and the correction highlight the problem that journalism by Twitter creates. In any professional news organisation, to write and publish a story is a process usually requiring oversight from at least one other person, such as an editor; to then remove or retract that story from the publication is a painful process, involving the consumption of lots of humble pie by the journalist. Yet although the BBC's rules on the use of social media are strict, they're not as strict as those on blogposts or stories that appear on its website, which must be checked by two other people before they are published.

In the age of social networks and smartphones, though, all three journalists were able to circumvent the normal editorial boundaries of their employers, and broadcast what would count as 'news' on a network owned by a company that competes directly with their own for attention and, in ITV's and the *Sun*'s case, for advertising revenue.

Why? Because, as journalists do, they felt the attraction of immediacy. News journalists live in thrall to 'the scoop'. They fear being beaten to it, and are eager to be first with it. Equally, organisations with an eye on their future audience feel obliged to go where that audience may be – and the plummeting figure for hours watching TV and reading newspapers, set against the rising time spent on social networks, shows where that is.

While Clay Shirky's 2009 observation about the Iran protests and Twitter – that one had to trade speed for accuracy – had been borne out, the only beneficiaries seemed to be Twitter and Facebook. The effect of the reporting had been to foment outrage, first to the initial reports and then to the follow-up corrections. Which side you took tended to depend on political affiliation, but it was hard to see that anyone was well served. Nothing about the incident enlightened voters about the desirability of one party's policies and politicians over another's, or how they might rule. But it distracted, and prompted people to write angry tweets and Facebook posts attacking the minister, or the protester, or the journalists, or some combination of them. Heated debate, but with no moderator and no timekeeper, and with no object but to continue in order to keep the social networks sated and advertisers satisfied. The question about the capacity of the hospital to treat children with pneumonia was overtaken by the reaction to a trio of tweets.

Those events bear comparison with another UK general election nearly thirty years earlier. In March 1992, the Labour opposition aired a TV party political broadcast (effectively, a political advert) seen by millions of people, about a five-year-old girl called Jennifer. The film showed her waiting for ear treatment on the NHS and becoming more and more miserable as she had to attend school untreated, while another unnamed girl with the same condition was seen quickly by private medicine.[71] (The intended message was that the Conservatives were failing to fund the NHS adequately. Some topics are evergreen.)

The morning after the broadcast, the right-wing *Daily Express* newspaper's front-page story was 'EXPOSED: Labour's Sick NHS Stunts'. It gave Jennifer's full family name, and said that her problems had actually been caused by administrative error, not underfunding. Labour was accused of using a child's discomfort for political gain and misleading the public. In response, Labour accused the right-wing press and the Conservatives (who turned out to have leaked the family name) of invading the family's privacy.

The row escalated over the coming days, with the hospital consultant and the child's parents and the editor of the *Daily Express* and the Conservative central command all being dragged in. In all, it lasted roughly a week.

Crucially, though, all of the discussion and information emerged through broadcasters and newspapers. All of the heat and misinformation was to a large extent controlled and balanced by the competing outputs. If you wanted to express an opinion about it, you could write a letter to the editor; it might be printed, though most weren't. Voters weren't enthralled by the politicians' behaviour, according to polls taken at the time. The row became known as The War of Jennifer's Ear, an ironic nod to the eighteenth-century War of Jenkins' Ear, a grinding Caribbean trade war between Britain and Spain.

What was radically different from the modern age is how the war ended: by fiat. Michael Heseltine, a centrist Tory politician, made a speech asking for the two sides to stop arguing about it, and concentrate instead on the other topics of the election. And they did.

Could that ever happen now? The original Jennifer, who is now a teacher in a town in Kent, south-east England, was struck by the comparison between her experience and that of the sick child on the hospital floor. The spread of the news had been quicker, she told me, but 'the act of weaponising a sick child is no less disturbing.' She hadn't been very aware of what was going on in 1992, but the consequences of that experience had been far-reaching for the whole family. (Her father was a Labour voter; her mother,

Conservative.) The effect was 'a ripple effect that will probably span my lifetime. There's a reason I don't publicly share my political views,' she said. 'I have no intention of "winning" any future elections!' Was the latest case just a social media forest fire, I suggested, which might avoid the family? 'I truly hope it will be just that and it dies down for the family quickly,' she said. 'Undoubtedly media has changed, but it seems politics hasn't moved with the times in the same way.'

In 1992, in a world of 'authoritative' media, the debate had principally been carried out between the political parties and the journalists, with the public as onlookers. In 2019, the public were able to pitch in their opinions – whether or not they were informed or useful. Most simply weren't, and the disinformation about the sick child meant that the topic of NHS funding, which would turn out to be crucial in just a few months when the coronavirus became a pandemic, was overlooked in favour of the social warming of Twitter and Facebook back-and-forths between people who wouldn't be persuaded of anything that the other side said.

Even worse, social networks have undermined people's trust in authoritative sources. Before social networks, people relied on news organisations to tell them what was true, and what was important – both important gatekeeping functions. Now, their faith even in the ones they had relied on could be swept away. On the morning of Tuesday 10 December, James Mitchinson, the editor of the *Evening Post*'s morning sibling, the *Yorkshire Post*, tweeted the response he had written to a reader. She had written to him saying she would stop buying the paper because of its outrageous lie about the child on the floor of the hospital: she had seen on Facebook that it was not true.[72]

While he walked his dog, Mitchinson composed a thoughtful response pointing out that the paper had not taken the mother's word for it, but had instead checked with the hospital, and received confirmation of the events from the hospital and its two most senior staff.

By contrast, he observed, the account that had posted the claim of 'fake news' had disappeared, while he and the paper were still there. There was no way to hold the author of the Facebook post to account: 'You don't even know if it is a real person, so why do you trust her claim over the newspaper you've taken for years, in good faith?' There was no source on the post apart from 'a good friend', which might make it sound human but these were words 'manufactured very carefully and cynically in order to mislead'. Mitchinson accepted that the woman who had complained might still not believe him; but, he implored her, 'Whatever you do, do not believe a stranger on social media who disappears into the night.'

On the day of the 2019 election two days later, the front page of the *Yorkshire Post* was dominated by a commentary by Mitchinson, expressing his frustration at people's inability to trust the media any more.

'For this newspaper the tipping point came when loyal, committed, lifelong readers began contacting the editor to advise him they would no longer be buying or reading the *Yorkshire Post* because they had seen a "Facebook post that proved we were lying,"' he wrote, continuing: 'Journalism that had been painstakingly fact-checked and meticulously sourced was being discredited by a digital disease that is hypnotic and alluring yet nothing more than a con trick propagated by conspiracy theorists.' The con trick wasn't about money: it was about 'insidiously diluting our ability to confidently discern truth from lies.'

Rather than calling on its (presumably older) readers not to believe stuff they read on Facebook, or perhaps just to dump it altogether, the paper called for the Electoral Commission to launch an inquiry into 'the conduct and content of this election campaign' because 'the people of this country must never again be asked to navigate a maelstrom of misinformation in order to decide who will govern them.'

The election came and went. The Electoral Commission made no announcement of any plans relating to conduct and content of the campaign.

'In terms of the media's relationship to the truth, journalism and the truth, I just think everything is getting worse,' Jesse Singal, the Brooklyn-based journalist we met in the section about the *Black Witch* controversy, told me. 'There's just this sort of viral mistruth problem, where stuff just gets passed around endlessly without fact-checking or quality control. Random people on social media are obviously going to pass round rumours, but seeing cracks start to appear in the edifice of journalism has been very discouraging.'

Twitter can also undercut journalists' ability to interact with their subjects. A reporter's 'biased' tweet about a topic (where the bias may be very much subject to interpretation) can be used by would-be interviewees as a pretext to refuse to engage with them.

'People are very sceptical when you complain about Twitter too much, because so few people are on it [in an absolute sense],' says Singal. 'But the people who matter are very much on Twitter. And it absolutely skews the way they do their day-to-day job in pernicious ways.'

He explains: 'If you're a journalist in a newsroom in America, every day you wake up and you log on to Twitter, and then you're on Twitter the rest of the workday. So whenever a controversy breaks, you immediately know how all your friends and colleagues and people ideologically aligned with you have been reacting to the controversy. There's no room to even develop your own opinion on anything, because we're social creatures.'

He cites an example he saw, of a story about harassment (which he declines to detail): 'a pseudo-MeToo story that so many people in journalism, people I know, thought was riddled with holes. But no one would say so online, because you would have been pilloried, seen as anti-woman, a sexual harassment apologist. But a lot of people who were strong progressive feminists really thought it was bad journalism. They were afraid, mostly, of social media, because they could see which way the winds were blowing.'

Journalism has never been a 'totally pure enterprise,' Singal told me. 'But there was for a while the sense of journalism as at least claiming to try to get to the truth, and it feels like those norms are slipping a little bit. It's been very depressing to watch. It's driven partly by the collapse of the economic structures of journalism, but is also driven by social media, frankly.'

The realisation that chasing social media eyeballs might also not be in the best interests of the BBC was dawning on the British public broadcaster. In spring 2020, a former executive, Richard Sambrook, was asked to look at how staff and presenters were using social media, and compare that against the BBC's current guidelines.[73] Those guidelines said that people should 'take particular care about maintaining our impartiality on social media' and that staff should 'avoid bringing the BBC into disrepute through their actions on social media'. Impartiality – taking neither side in a debate, and treating opposing voices as having a valid right to be heard – is a dearly cherished quality for BBC journalists, who exist amid a press that is frequently partial. In particular, the BBC social media guidelines noted that 'each atomised piece of content should be judged on its own merit' and warned that 'we should be aware of the risks that short headlines, tweets or the pressure to create shareable content pose' to impartiality.

But might the problem simply be how attractive Twitter had become to some journalists? 'The way social media has developed in recent times – particularly Twitter – has become adversarial, more argumentative, more combative, more polarised and sometimes toxic,' acknowledged David Jordan, the BBC's director of editorial and policy standards, in a hearing at the House of Lords in July 2020. 'It can suck people in. The immediacy of it can be alluring. The live dynamics of it can be seductive to some people.'[74]

It's a long way from Janis Krums taking a photo of a plane in the Hudson. Yet the same instinct – 'that's pretty newsworthy' – is driving people. Stuff I care about, stuff I want to pass on – even if it makes everyone a little more confused.

8

DEMOCRACY AT RISK: WHY SOCIAL MEDIA UNDERMINES ELECTIONS

> *The peculiar evil of silencing the expression of an opinion is . . . If the opinion is right, they are deprived of the opportunity of exchanging error for truth; if wrong, they lose, what is almost as great a benefit, the clearer perception and livelier impression of truth, produced by its collision with error.* – John Stuart Mill, *On Liberty*

While the effects of social warming on our attitudes to truth, politicians and media are all important, none matters as much as what can happen to democracy under the influence of an attention-hungry algorithm set to make outrage go viral. Democracy is fragile: the capture of the levers of state by small groups obsessed with authority can lead to the collapse of democratic institutions and the rule of law.

Based on what we've seen so far, it might seem that the lower the penetration of smartphones and social networks, the safer the rule of law should be. But the example of Myanmar, and the pre-smartphone experiences of Rwanda and Bosnia, show that isn't necessarily the case. Tensions can exist for decades or centuries until they find a release, and like a volcano erupting, the effects can be disastrous for those who are in its path.

When the Arab Spring led to the overthrow of authoritarians, and social media was a key underlying system that helped people to organise and find like-minded malcontents, the praise was lavish, even while the social networks mumbled and pretended to be flattered. 'What I like to think of services like Twitter and other services is that it's kind of a supporting role,' Twitter's then CEO Biz Stone said in February 2011. 'We're there to facilitate and foster and to accelerate those folks' missions.'[1] Zuckerberg too suggested the same: in 2010, as the company reached the milestone of 500 million users, he said that 'The power of democracy in these systems, is that when you give everyone a voice and give people power, the system usually ends up in a really good place, so what we view our role as is giving people that power.'[2]

But if social media isn't widely used, and those people don't have that voice, how could it affect a democracy? Or, on the other hand, if it's pervasive, won't the natural balancing of the 'marketplace of ideas', the push and pull of the political debate, mean that truth and accuracy will win out, as John Stuart Mill suggested, leading voters to pick the person who is telling the truth? Won't making the world more open and connected mean that everyone benefits overall, as Zuckerberg keeps insisting?

The guitar of folk singer Woody Guthrie carried the label 'This machine kills fascists', but the ultimate weapon for that was meant to be social media. Fred Turner, professor of communication at Stanford University, points out that after the Second World War, a favoured theory for how Hitler and Mussolini had come to power (and how fascism in America had had a brief moment in the sunlight in 1938) was the power of mass media.[3] The ability to push a few voices, or just one voice, to a huge audience was a dry run for the one-to-many form of fascism. Therefore it stood to reason – didn't it? – that having many sources of information able to reach everyone individually would defeat any attempt to corral people's thinking down one path. What was needed to forestall fascism and authoritarianism was not mass media, but *personal* media. The promise of Facebook and Twitter, Turner observed, was that 'social media will allow us to present our authentic selves to one another. To "connect", and so by implication form an egalitarian, even potentially anti-authoritarian, solidarity.'

There even seems to be some evidence that that's correct. Steven Wilson, assistant professor of political science at the University of Nevada, has explored the effect of the internet on authoritarian regimes for years. Looking back at the effects of the internet over the past twenty years, he told me that 'obviously the utopian view of things didn't end up very utopian. But there is a decent amount of evidence that social media and the internet did have a lot of the predicted effects.' Between 2000 and 2016, the rate at which dictators were overthrown was twice as high as the previous thirty years – a period that included the fall of the Soviet Union. And Nick Clegg, Facebook's vice-president for global affairs and communications – in effect, its top PR man – said in June 2020 that 'Political polarisation has fallen in many countries with high internet use.'[4] (This seems to be a reference to a study published that January, led by Jesse Shapiro at Brown University.[5] That looked at nine countries – the US, Canada, New Zealand, Britain, Germany, Sweden, Norway, Switzerland and Australia – and in fact found that any fall in polarisation was only statistically significant in Sweden and Germany.)

What Clegg didn't talk about is a different, bigger problem: the way that social media is used more and more to interfere in and to influence elections, which are the clearest expression of democratic freedoms. It doesn't even have to be heavily used inside the country to make a difference.

Elections are the perfect breeding ground for the effects that we see again and again from the combination of always-available smartphones and social networks: the most extreme voices that can shout loudest and longest drown out the others, and drive the debate towards dichotomy. There can be no middle ground between zero and one. You're a supporter, or you're an opponent. Mix in the amplification, the bots (centrally controlled networks of inauthentic accounts), and the disinhibition effects of social networks – you're not worried what you say, because how will it come back on you? – plus the pervasiveness of smartphones, and the shape of the problem starts to become clear.

The places where this should worry us more are the countries where democracy is weakest – where democratic and media institutions are already under threat or compromised – and electorates are least educated or informed because access to accurate information is expensive.

Ethiopia, in north-east Africa, has a population of around 110 million. It's one of the world's poorest countries, where the affordability of the internet – measured as cost against average per-capita income – is near the bottom of global rankings, as are internet penetration rates. In 2010, only Myanmar and a handful of countries were worse, but Ethiopia fell behind as internet-enabled phones took off in the Asian country after 2010. Like Myanmar, however, ethnic tensions between the majority Oromo group and others infect Ethiopia's politics.

The idea that social media could be exploited to influence elections there might thus seem far-fetched. Only around one-fifth of the population, or about 22 million people, have any internet access, and only about 5 percent, or 6.2 million, use social media – almost all of that Facebook. (The now-familiar phrase that 'many citizens consider Facebook to be the internet' applies here too.) How could such a comparatively small number affect such a populous democracy, even one as fragile as Ethiopia's? A measure of how often average people use social media to organise real-world political action also puts Ethiopia near the bottom of the world rankings.[6] But when in 2016 videos showing violence by the military and police were put online, they sparked the sort of outrage seen previously during the Arab Spring.[7]

Yet if comparatively small numbers of people spread messages on Facebook and Twitter, does that matter to the wider population?

'It's a sort of perfect testbed for this,' Wilson told me. 'Here you have a formerly authoritarian country that's had a recent democratic transition. Maybe five percent are on social media. And yet we actually can tease out major effects of social media and the internet on its politics.'

Wilson says that a key moment came when the opposition was able to organise and push through the democratic transition in 2018 because they could organise online. For years before, the Ethiopian government through its control of the telecoms system had blocked and filtered sites, particularly blogs critical of the regime.

Even if social media didn't look important to outsiders, the government certainly acted as though it was. Documents leaked to activists in the diaspora in January 2018 showed that the government at the time was paying people to promote the regime on social media.[8] A number of people, including two publishers of Ethiopian news sites, were being paid the equivalent of half a year's average salary (US$300) for blogposts or Facebook comments and posts backing the regime. The resulting protests forced the prime minister at the time, Hailemariam Desalegn, to resign; he was replaced in April 2018 by Abiy Ahmed, who had been in charge of digital surveillance under Desalegn.

Abiy certainly seemed to want positive change. He freed hundreds of political prisoners, legalised opposition groups, appointed women to his cabinet and removed a number of corrupt officials. He sought to end the war with the country's northern neighbour Eritrea, begun in 1998 but in its eighteenth year of stalemate; he had fought in it as a child soldier. He also became the first Ethiopian leader with an active social media presence, as internet censorship was lifted.

Abiy's progressive moves were celebrated by many activists in the diaspora, including one called Jawar Mohammed. He had been living in exile in Minnesota, which has a large Ethiopian population, for some years and had helped organise the 2016 protests through his Facebook Page and personal media network, which included a TV channel and news site aimed at the majority ethnic Oromo group, who live mainly in the south of the country. 'He built a million Facebook followers there and returned, kind of triumphantly, despite the very low social media access [in Ethiopia],' Wilson explains.

Although he was initially supportive of Abiy, Jawar's disagreements with the leader began to grow. By mid-2019, the country's ethnic tensions were also beginning to heat up, warmed by rumours spread online that were then repeated offline. Abiy responded as he had in his previous job by imposing internet blackouts – a move that Human Rights Watch later said had become routine since he came to power.[9] After the assassination of five government officials in a military coup attempt in August 2019, Abiy commented that he might cut internet access off 'forever'; the internet was useful but, he said, was 'neither water nor air'.[10]

By that point, the easing of internet controls meant Ethiopia had its own collections of YouTube and Facebook broadcasters pushing inflammatory content about ethnic groups or conspiracy theories about events. 'They are mostly diaspora-based monologuists broadcasting from their living rooms, complaining vigorously about the Ethiopian government, and attacking each other,' observed Dr Endalkachew Chala, writing at Advox, a global activist network.[11] The comments under the videos and posts were packed with people insulting each other, he noted. Good business for Facebook; social warming for Ethiopia, echoing in its own way the crisis that had swept over Myanmar a couple of years earlier.

Among the conspiracy theories being pushed was one that Abiy had instigated the military coup 'in order to crush Amhara [another ethnic group] nationalism.'

A number of concerned activists started a campaign to flag such disinformation and related hate speech to Facebook and to YouTube. In some cases, the videos and posts were in Amharic, one of the ethnic languages.

They were dismayed to find that almost all of the content stayed up, including the edited version of a YouTube video showing an extremist threatening to kill Amhara families living in Oromia. It was still circulating on Facebook months later, to the frustration of the activists who, Chala reported, 'wonder if Facebook cares about the damage it might be causing in Ethiopia'.

Ethnic tensions, conspiracy theories, non-English languages, moderator indifference: some elements of social warming become so familiar with the retelling that it's only the location and fine details that change. The same supranational detachment that lets social media companies ignore or slow-walk government requests for data about citizens also works against those citizens when applied at scale to dangerous, socially inflammatory content.

But how does social media affect politics in a country where most people aren't online and aren't on it? Wilson thinks the most important factor is 'the tip of the iceberg effect'. He explained: 'Elites and pseudo-elites can take advantage of the organisational capacity of the internet, even if a large bulk of the people in the country aren't on the internet or using social media actively. There can still be a huge multiplicative effect for them.'

Early in October 2019, Abiy was awarded the Nobel Peace Prize for his efforts in ending the war with Eritrea. Less than two weeks later, the country was convulsed with unrest after Jawar posted on Facebook that the government was preparing to arrest or kill him. His social media leverage was demonstrated when a huge crowd turned up at his house. The situation quickly turned violent.

Thousands protested against Abiy, leaving more than eighty dead. After two days, Jawar called for the protests to stop – and they did.

Receiving his prize at the Nobel ceremony in December 2019, Abiy warned that 'We should avoid the path of extremism and division, powered by politics of exclusion ... The evangelists of hate and division are wreaking havoc in our society using social media,' and called for citizens to 'neutralise the toxin of hatred.'[12] His solution was extreme, including, from January 2020, shutting down the internet in three parts of Oromia where the military was fighting insurgents.

Certainly, the government in Ethiopia, and its opponents, behave as though they see social media as crucial to their success. Photos are posted every day on Abiy's Facebook and Twitter accounts (with, respectively, more than 1.2 million and 350,000 followers). Jawar threatened to mobilise his followers if he felt the elections, due towards the end of 2020, were rigged. And both were clear that the most important battleground was social media.

'There's no threshold now with social media,' says Wilson. 'You used to need to own a major media company to reach millions of people. Now you need a Twitter account, and the more extreme you get, the more people you'll reach with your message. And all the moderates in the middle will have this "Well, I don't necessarily buy it, but if it is true ..."'

The key problem, Wilson suggests, is the lack of moderating forces online. 'There's normally a cost of entry on some level into the public debate.' But online, anyone can participate. That doesn't create a fairer or more reasonable world; it creates a world with many more people shouting to make themselves heard, in some cases honestly, in others at the behest of larger groups. 'We just have enormous amounts of extremism, and fake news, and propaganda, and people just pushing blatant lies online,' Wilson says.

* * *

The Philippines is one of the most connected countries in the world, and its population of 100 million people has form for using technology to achieve political change: in January 2001, thousands protested against the refusal of the senators to investigate evidence that could impeach president Joseph Estrada. The movement was organised by a waterfall of text messages calling on people to gather at a famous shrine. Estrada left office four days later.

By 2012, the internet was rising quickly as a source of news there.[13] Though almost 90 percent got some information from TV, the second-biggest source was the internet, cited by 45 percent – while the radio, newspapers and magazines were seeing rapid year-on-year falls. By 2016, 97 percent of internet users in the country were on Facebook ('What about the other three percent?' said Zuckerberg when told the figure), helped by the launch there in October 2013 of Free Basics – Facebook's program that persuaded partnering mobile networks to give customers unmetered access to about twenty sites, including Facebook, Wikipedia, BBC News and a few others.[14] On the other sites, comments and pictures were removed or shrunk down; the only way to see them full size was to spend money on a paid plan. Naturally, people stuck to the limited number of sites, and particularly to Facebook.

Zuckerberg was delighted with the free plan's success; a few months after the launch he called it 'a home run' and said it was 'going really well'.[15] By 2016, more than half of people had mobile broadband (against 10 percent with fixed broadband), and 80 percent of them were on the slowest, cheapest data plans.[16] Yet by 2017, Filipinos spent the most time daily on social media in the world – on average more than four and a quarter hours.[17] That put them well ahead of Brazil (3h 45m) or Argentina (3h 32m). But that also meant people were vulnerable to the manipulation that social networks enable. By 2018, two-thirds of the population were on Facebook; the rest didn't have internet access.

Effectively, Free Basics walls off the rest of the internet. For the mobile carriers who actually fund it, the scheme is a great way to hang on to customers; for Facebook, it's a great way to keep people on the site, to be shown adverts. Hardly anybody would want to incur the cost of going to a news site that wasn't part of the free package.

Yet not everyone thought it a 'home run': a critical report in 2017 by Global Voices called Free Basics a form of 'digital colonialism', and pointed out that the system 'features an imbalance of sites and services' and 'does not allow users to browse the open internet'.[18] (India banned the scheme in 2016 for precisely that reason.) Equally concerning was that the scheme didn't work as Facebook intended, to give access to people who otherwise wouldn't be able to afford a data connection. Instead, about nine in ten already paid for data, but used the Free Basics sites when they ran out of credit. And when that happened, all they could see on news sites were headlines and photo captions – not the actual articles.[19]

Of course, that creates the perfect conditions for social warming: a single site with minimal external oversight on which everyone in a highly contested electoral campaign would try to make themselves heard. Politics in the Philippines has never been clean or respectful. Facebook and its walled garden held the population in and provided a pressure cooker that dramatically raised the electoral temperature.

The message about politics and activism that those trying to push extremist agendas had learned in the years following Gamergate was simple: a successful online campaign didn't have to be built around truth. What mattered was having a group of dedicated people who could marshal talking points and manipulate the attention-seeking algorithm in their favour. In online campaigns, having a few people who are willing to give almost all their time to something is far more valuable than having a large group of mildly interested people. That tends to apply in a clash of views:

the most determined will prevail, partly because the algorithm will show their content to the undecided. If you can also get fake accounts to amplify that content, the algorithm will follow suit; the machine doesn't know it's untrue.

One of the candidates in the presidential election in early 2016 was the tough-on-crime 71-year-old mayor of Davao City. His name was Rodrigo Duterte, and his age belied his quick understanding of how social media could be used to connect directly with people and spread his populist, brutalist message. Nuance didn't matter; nor did balance. Duterte hired a social media strategy team who in turn built up a formidable affiliation of hundreds of bloggers and Facebook personalities willing to echo his policies, tell outright lies and smear his opponents.

To Facebook's News Feed, though, anything generated by someone on Facebook that someone else wants to read is OK. That includes news sites that are actually just manufactured content. In March, a campaign Page on Facebook posted an image of Pope Francis with a long caption apparently praising Duterte, ending with the words 'I admire his honesty' – a comment sure to have a huge impact on the heavily Catholic electorate. Except it wasn't true, as the Catholic Bishops' Conference of the Philippines tried to point out.[20] 'This statement from the Pope IS NOT TRUE,' they wrote on their own Facebook Page. 'It came from a satire piece and is a fake.' (The use of 'satire' to evade Facebook's fact-checkers was just coming into view.) The comments below the bishops' denunciation were a mudwrestling match of pro- and anti-Duterte supporters.

That was far from all. Notably, the organisational methods that succeeded weren't the ones used in 2001, where text messages found those individuals who were prepared to take physical action. Instead, there were virtual calls to action that required nothing more than clicking on a link on a screen (and sometimes even less; just reading the content). Nobody had to turn up to protest anywhere, which allowed the fake nature of many of the accounts behind the loudest output to remain hidden. A faked sex tape showing Duterte's female opponent began circulating. Facebook became the conduit for all of it. Certainly, people still had to vote. But that's the minimum act in a democratic society.

Facebook itself acknowledged Duterte's effectiveness during the election. 'The most talked-about person in the Philippines election,' said a press release as the rivals struggled for primacy. Elizabeth Hernandez, Facebook's head of Asia-Pacific public policy, said proudly that 'Candidates are using Facebook to reach citizens directly and personally. They are having unfiltered conversations about important public policy issues and getting honest feedback from voters.' Yet the 'talked-about' measure she quoted wasn't how many 'fans' Duterte had (he ranked third, of the five rivals), how many posts he'd made (he was fifth) or how many comments his posts garnered (he was fourth), which would have counted as 'feedback from voters'.[21] What Facebook focused on was how much people shared content about him, a measure on which he was far ahead of his rivals.

The document perfectly illustrates how being able to measure something doesn't necessarily mean it's the correct thing to measure. Duterte was being discussed, certainly, but might it have been because he was notorious for turning a blind eye to extrajudicial killings by police? Or was the discussion being distorted by bots and inflammatory content? It's like measuring success in shipping by how many days you can sail through the Arctic free of ice: if the numbers are growing, you may be missing the wider picture.

The measure wasn't even necessarily truthful: during the election, just twenty-six fake Facebook accounts generated content that reached three million people, according to a subsequent investigation by Maria Ressa, a Filipino former CNN investigative journalist who had set up an independent news site there called Rappler.[22]

Duterte was elected in May 2016 – somewhat to his surprise. But he realised that social media had put him into power, and so he brought it into his governance. His inauguration was streamed on Facebook Live; the press was banned from the event.

The online attacks and smears continued after the election. Filipinos alarmed by the divisive, inaccurate content and the threats being spread by those backing the Duterte campaign tried to warn Facebook about the abuse. Ressa met senior Facebook executives in August 2016 and pointed to the personal attacks and fake news that Duterte supporters were posting on the site, yet which were allowed to remain. She later said that she felt as though the executives weren't familiar with their own site: Facebook took months to take any action at all, while Ressa was subjected to thousands of online attacks after writing about the misinformation scheme in a three-part series, beginning in October 2016, titled 'Propaganda war: Weaponizing the internet'.[23] Ressa complained again and again to Facebook, repeatedly encountering polite indifference: Facebook always saw it as a matter of freedom of speech. 'When Mark Zuckerberg sticks to a position that it's OK to lie, that's the fundamental flaw,' Ressa said in June 2020.[24] 'If you distribute a lie further and faster than facts, you create a society of lies.'

But it's not in Facebook's interests to calm down political discussion, to rein in the wilder claims, or even to protect the democratic process by ensuring that the discussions online reflect real voters and their views. That wouldn't bring people onto the site, and given that the wildest people tend both to post the most and to draw the most attention, reducing their visibility would be counterproductive – or 'anti-growth'. Even if there's not much money to be made from showing adverts to people in any particular country – Myanmar and the Philippines are around the 150th and 120th in the world for per-capita GDP – it costs Facebook nothing to add an extra user, and wrests them away from potential rivals. Facebook doesn't care about truth in elections. It cares about voters and non-voters using its site.

Nor does destabilisation of democracy always need the amplification of Facebook. Sometimes algorithms aren't needed because people will do a lot of the work themselves, aided by our natural tendency to pass on the outrageous, the attention-grabbing, the conspiratorial, because that is in our nature. The system then helps by spreading the false, the flawed, the attention-grabbing but fake.

Such was the case with WhatsApp in Brazil. The app has more than 120 million users in a population of 210 million. 'I think that every family in Brazil has a WhatsApp group,' Benjamin Junge, an anthropology professor who visits the Federal University of Pernambuco in Brazil, told the Vox news site ahead of the 2018 elections.[25] Families, friends, students, teachers, churches – everyone, said Junge, used it: 'I can't even really imagine what this election would look like without WhatsApp.' He felt sure that the messaging app was an important conduit for voter information.

But how could WhatsApp, which needs a phone with a data connection (whether Wi-Fi or mobile) have become pervasive? Many people in poorer areas couldn't afford data for their phone; Brazil is one of the five most expensive countries for data plans.

You may have guessed: Facebook, which owns WhatsApp, had negotiated zero-rating for data used inside it. Around three quarters of Brazilian internet users had such zero-rated plans in 2016, giving them free use of apps including WhatsApp, Facebook and (with some operators) Twitter.[26] Anyone with an internet connection at home or work could push YouTube content onto WhatsApp groups, where the individual videos could be shared endlessly. But anyone inclined to fact-check what they saw in a video would have no direct access to a search engine without a pricey data connection. Furthermore, as in India and other countries, many of those targeted by misleading or faked political messages had low literacy levels, and Brazil's TV networks were not always interested in high-quality journalism. It was fertile ground for political disinformation and misinformation to spread.

One spring that watered the ground was YouTube itself. An investigation by the *New York Times* in August 2019 found that right-wing sources had pushed conspiratorial content during the election that would trigger viewers' feelings of anger and fear, which they would then spread further.[27] YouTube's recommendation algorithm began showing people more and more extreme political content; its takeover of their attention subsequently played a crucial role in getting a number of candidates elected. After all, just like Facebook or Twitter, YouTube's algorithm isn't tuned to decide whether what someone says is true, or even grounded in truth. All it sees, *Breakout*-style, is the brick wall of viewing time to be broken down by aiming videos at people.

A linked part of the *New York Times* investigation confirmed that despite having no algorithmic reinforcement, WhatsApp had been an important channel for the amplification of misinformation during the 2018 Brazilian elections.[28] Quite how important was demonstrated by an elegant study carried out by a group led by Virgílio Almeida, at the Federal University of Minas Gerais in Belo Horizonte. In late 2017, they collected three weeks' worth of messages from eighty-one public WhatsApp groups based in Brazil that had nearly 7,000 users in total.[29] With presidential elections due the following year, two-thirds of the groups were taken up with discussion of politics, and generated four times more messages than the others. Notably, one-fifth of what was posted wasn't text, but audio clips, pictures and video. Zipf's Law applied: 'most of the content is posted by few users that seem to dominate the discussion quantitatively,' Almeida's group found.

In a country where 44 percent of voters used WhatsApp to get political and electoral information, that lack of moderation in concert with Zipf's Law is concerning, as it means that a tiny minority dominate the discussion, while being completely beyond the reach of normal checks and balances expected in a functioning democracy. If one party or candidate overwhelms the input to enough groups, which then spill over into others where voters cannot or will not check the message for its truthfulness, the process of democracy is subverted. There's a stark resemblance to the catastrophe that overtook Myanmar, where Facebook's moderators (once there was more than one) were still handicapped by trying to read Zawgyi-based content that their Unicode-based computers could not render correctly. The only moderation on WhatsApp is by the individual administrators of groups, although the company can – and does – take action against accounts that it deems suspicious (though what makes them qualify as such is never discussed).

Almeida later found that users in Brazilian WhatsApp groups were uploading one video for every fourteen written messages, a remarkably high proportion. YouTube was linked to ten times more often than Facebook, and more than any other site.

The election campaign run by Jair Bolsonaro in late 2018 was also accompanied by a tornado of social media manipulation – and by a feedback loop of misinformation shuttling between YouTube and WhatsApp. The echo chamber of WhatsApp groups, the lack of moderation, the dominance of a few louder voices, and the challenge for data-poor users to check claims (even if they wanted to) was just as effective as an amplification algorithm. Poorer and less literate Brazilians were the ones who shifted to vote for Bolsonaro, an ex-soldier whose racist, misogynist and homophobic comments, and admiration for the country's former dictatorship (which ended only in 1985) were shocking to so many voters who opposed him.

A team at the Berkman Klein Center at Harvard University also found something else about YouTube's content in Brazil: the algorithm was increasingly recommending right-wing videos to people after they had watched unrelated content. In effect, by paying attention to what people in the country were watching, the system concluded that others should try it too. The algorithm had no notion of political impartiality, or equal airtime for other politicians; nor awareness that it was creating a feedback loop. Max Fisher and Amanda Taub, reporting for the *New York Times*, noted that far-right politicians there had seen their audience numbers 'explode'. (A number of far-right Brazilian YouTube personalities were also elected to various positions in the 2018 elections.)[30] YouTube disagreed with the researchers, telling the *New York Times* that 'internal data contradicted their findings' – but then, as so often, refused to share the data, or any statistics that could have shed light on what was happening. The commitment of Google and Facebook to 'transparency' tends only to apply to others. On their own products and data, they are almost uniformly opaque.

The big concern is that if people from one political side or the other learn how to flood YouTube and put their thumb on the scales of the algorithm, there could be a domino effect among viewers, who will be led more and more towards extreme political positions. YouTube could claim that it was not interfering actively in the democratic process. But passively? That would be a different matter.

When I asked YouTube, it responded that 'Anyone can choose to post videos on YouTube without interference and regardless of political affiliation, as long as they follow our Community Guidelines. In addition, our recommendation systems are not designed to filter or demote videos or channels based on specific political perspectives.'

Which evades the point, of course.

* * *

Bolsonaro's election wasn't comparable to the narrow-squeeze victory of Donald Trump in 2016; Bolsonaro won by 55–45 percent in the final stage. But earlier in the contest he had trailed the former president, Lula da Silva, by a wide margin[31] – until Lula was declared ineligible because he was in jail.[32] The momentum that Bolsonaro gained after that point indicated reinforcement of his message. The campaign was particularly notable for the sheer volume of misinformation aimed at the left-wing opponent, Fernando Haddad. In mid-2018, Fabrício Benevenuto, a computer science professor studying social media at the Federal University of Minas Gerais, created a 'WhatsApp Monitor' to observe content flowing through a number of public groups.[33] It offered a concerning view of reams of untruth. Benevenuto called it 'a look through the crack, a keyhole'.

A week before the election, three Brazilian academics wrote in the *New York Times* that WhatsApp should make changes 'to reduce the poisoning of Brazilian political life.'[34] Monitoring nearly 350 public discussion groups between mid-August and early October, they had found that only 4 of the 50 most-shared images in them weren't faked or manipulated in some way. Many others were real but were accompanied by captions or text that used conspiracy theories or wild claims (such as one suggesting that Lula and 'FHC' – Fernando Cardoso, the Brazilian president between 1995 and 2002 – had met in bars to plan bank robberies resulting in deaths).

'The problem of false news in Brazil transcends ideological divides,' the researchers pointed out, noting that anti-Bolsonaro content had been shared widely too. To prevent the problem, they suggested WhatsApp should severely limit content forwarding, capping it at five groups (as had been done in India after a series of WhatsApp-inspired lynchings) and limit the size of new groups. They said the company had told them there was 'not enough time' to do that before the election, even when the researchers pointed out that doing the same in India had only taken a few days.

'Brazilians should not be casting their votes on the basis of false or distorted information,' they complained in the article, saying the measures were needed 'to stop the spread of fake news and dangerous rumours ahead of a critical election.'

By going to the *New York Times*, one of the two biggest names in American newspapers, the researchers were trying to get their frustration with WhatsApp heard not only by Facebook in the run-up to the American midterm elections (which were also only weeks away), but also by American legislators considering measures against social media companies that seemed too big for regulation, and indifferent to the social damage they might be causing.

Days before the Brazilian election, Facebook terminated 100,000 WhatsApp accounts run by marketing agencies apparently paid by pro-Bolsonaro groups, and restricted the ability to forward content to other groups, to a maximum of twenty – though that would still allow a single message to reach the entire population, based on WhatsApp's average group size (which happens to be six).[35] 'There's no question WhatsApp is part of the electoral conversation,' said Victoria Grand, WhatsApp's vice-president of policy and communications. The trouble was, as Facebook implicitly seemed to recognise by its creation of a 'war room' in Menlo Park to try to watch what was going on there, the conversation was shot through with untruths. A post-election study published in October using WhatsApp Monitor found that nearly half of the information pushed by right-wing sources was false, including conspiracy theories about election meddling; for left-wing sources, the figure was just 3 percent.[36]

As in the Philippines, the social media assault didn't stop after the election. In April 2019, the Brazilian minister of education pushed through a 30 percent cut to university budgets, claiming as justification that they were a 'shambles' full of 'naked people'. This was quickly fact-checked by journalists as untrue, but that didn't stop WhatsApp being flooded with scores of pictures of university-age students (and non-students), none fully clothed: within twenty-four hours, the number of such pictures increased 100-fold, according to the researcher's keyhole view of public groups. On Facebook, posts with the misleading content had been shared more than 100,000 times. The pictures were old, out of context, misleading; that didn't stem their spread.

A panel of politicians convened after the election to look at what they called the 'industry of fake news that is shaking Brazilian democracy,' and began to focus on those close to Bolsonaro as the axis of that industry. In June 2020, the same researchers who had the keyhole into WhatsApp groups began analysing the sources of online attacks against those critical of Bolsonaro, amid the coronavirus epidemic (to which his response had been first to deny it, then downplay it, then ignore it, then catch it). Data about the sources of the online attacks pointed to people close to the president – his children, aides, businessmen, and bloggers. One politician who had fallen out with Bolsonaro, and then seen faked pornographic images of herself posted on groups, testified that she had discovered that the president's sons were involved in an organisation that used disinformation to try to take down those seen as opponents.[37] 'This is to kill you morally,' she said. Another of Bolsonaro's political opponents commented, 'If he didn't have fake news on social media, he wouldn't survive.'

The effect is evident here. Political forces can use social media to spread lies and smears, which is far easier than squashing them. People's adaptation to available channels – getting content from YouTube and then spreading it to WhatsApp, where it can be viewed for free – shows how eager we are to use any channel we can to gossip. Once more, the analogy with cars is apt: they're so fabulously convenient, and you can go anywhere in them. It's only after some years that the smog covering the city becomes evident, and breathing problems become widespread. If political discourse becomes hopelessly polluted by social media content, the problem isn't necessarily with politics, but with what social media does to it. If social media content and then YouTube's attention-oriented algorithm start to influence what happens in politics, and how politicians are viewed, it becomes important to know that what's being shared accurately reflects what has happened and what is being promised.

WhatsApp presents a particular problem: unlike YouTube, which is at least generally visible, there is no central moderation for WhatsApp, and even with limits on forwarding a message, the ability for a single user to belong to multiple groups means that oversight becomes nearly impossible. WhatsApp's engineering team has a helicopter view of what is going on, but they're hovering over a city watching people go into office buildings or houses: there's no way to see precisely what is going on inside.

Even so, WhatsApp's vice-president Chris Daniels felt compelled to write an op-ed piece for one of Brazil's largest papers just ahead of the election, insisting that 'more than 90 percent of messages sent on WhatsApp in Brazil are between two people' and that 'most groups have around six people – a conversation as personal and private as in a living room.'[38] When you connect more than a billion people, 'you will see all the good that humanity can do, but also some abuse,' he commented – arguably getting the proportions on good and bad woefully wrong. 'We have a responsibility to amplify the good and mitigate the bad,' he went on.

With the 256-person limit on group size, a single person would need to create about 4,000 groups to reach a million people, he pointed out. (Though of course the problem doesn't come from single actors, but from viral spread.) Then he admitted that hundreds of thousands of spam accounts had been removed, that

WhatsApp was working with fact-checkers and that 'we are raising awareness of the problem' with a public campaign called 'Share fact not rumours', intended to reach fifty million people.

It hardly sounded as though he was confident that WhatsApp was a neutral factor in the election.

'The conditions that allowed fake news to thrive in Brazil exist in many Latin American, African and Asian countries,' warned Luca Belli, professor of Internet Governance and Regulation at the school of law in Fundação Getulio Vargas. He points to the limited nature of the internet available to so many (because of the zero-rating for data to some sites) as a key problem.[39] 'The "rest of the internet" is precisely where Brazilians might have verified the political news sent to them on WhatsApp during the 2018 election. Essentially, fact-checking is too expensive for the average Brazilian.' The extra irony is that the video carrying YouTube content requires hundreds of times more data to transmit than a simple text-based page such as Wikipedia, or one from a fact-checking site.

In June 2020, the Atlantic Council, an American think tank that focuses on international affairs, announced the discovery by its Digital Forensic Research Laboratory (DFRLab) of a sophisticated Facebook operation run by a PR company based in Tunisia trying to influence elections in Africa.[40] 'Operation Carthage', as the scheme was dubbed, had created Facebook Pages that pretended to be legitimate news outlets and which targeted voters in ten countries, including Togo, Côte d'Ivoire, Tunisia, Senegal, Guinea, Madagascar and Chad. The operation had certainly been good at attracting attention: nearly four million Facebook accounts in the relevant countries followed the fake Pages, and nearly 200,000 on Instagram.

This wasn't the first time. In May and August 2019, Facebook had taken down accounts created by Israeli, Egyptian and

Emirates-based PR companies pretending to be news outlets, which amassed millions of followers for their misleading, and unannounced, 'influence' campaigns.[41] The companies promoted themselves as offering 'winning campaigns' for political influence – a glimpse of how social networks are now seen as a fundamental element in raising the electoral temperature: again and again, the Pages identified and removed by Facebook were trying to stoke outrage with distorted stories about terrorist funding, or dismissive memes about political candidates: 'Make Nigeria Worse Again' was the title of one Page aimed at a presidential candidate there in 2018. The DFRLab also found the same company running a Facebook Page that seemed to be in favour of the same candidate – and reasoned that its aim was to help identify his supporters so they could be targeted with opposing content. A Page that said it was exposing fake news in Mali claimed to be written by a group of students based there, except that Facebook's on-page tools showed that the controllers were based in Portugal and Senegal.

That might sound trivial. But the amounts being spent show it wasn't. The Israeli company spent nearly a million US dollars on Facebook adverts, paid in multiple currencies – US dollars, Israeli shekels, Brazilian reals. 'Conducting influence campaigns represents a lucrative and in-demand service offered by unscrupulous marketing firms,' the DFRLab observed.

The influence attempts don't even have to occur during an election period to erode democracy.

Which brings us to the gigantic 'influence operation' run by the British PR company Bell Pottinger in South Africa, which began in mid-2016. That involved the creation of more than a hundred Twitter accounts and of Facebook Pages that tried to foment a totally confected anger at white owners of capital in the country (using the divisive phrase 'white monopoly capital') while boosting the image of the Indian billionaire Gupta brothers, who were paying for the campaign, and deeply involved in suspected corrupt dealings with the then president, Jacob Zuma.

Twitter is widely used in South Africa: at that time there were nearly eight million users in a population of fifty-six million, and nearly fifteen million Facebook users.[42] The fake accounts retweeted each other, and the Facebook Pages gave the false impression of popularity and importance.

Racial tensions rose as elections approached, and the network of social media accounts, hand-in-hand with major media outlets under the Guptas' influence, sought to discredit and remove the finance minister (who had closed the Guptas' bank accounts), and to block a report into how Zuma had corruptly used state funding for his own ends.

As a strategy, the intertwined use of social media and mass media is an almost sure-fire winner for perpetrators. For those who believe that social media is an unfiltered source of truth, it offers a view of 'what the mass media won't show you'. If the mass media then picks up what's being said on social media, that becomes a vindication and validation of the story; even if some part of the mass media thoroughly debunks the story, a few people will always remain persuaded that it was true.

'Understanding how such campaigns work is vital for safeguarding against their underhanded influence today and for guarding against them tomorrow if we are to see our fragile democracy grow and thrive,' said a report into the network by the African Network of Centers for Investigative Reporting (ANCIR).[43] There was some solace: when Bell Pottinger's role in the scandal emerged in 2017, it lost £8 million worth of client business in two days, and filed for bankruptcy not long after. But the damage was done.

'It was the first large-scale fake news propaganda war in South Africa, and it has adversely affected the country's politics and economy,' said ANCIR in an article for the South African newspaper the *Sunday Times*. 'What the Guptas' multimillion-rand spin campaign successfully did was conflate the fight for economic justice with its own campaign to accumulate wealth.'

But the episode also demonstrated that even in a country with relatively low internet penetration, social media is nevertheless now seen as an essential tool in any attempt to influence public opinion. It's an expectation. Rousing fake anger against a government minister through a medium that only a small well-off proportion of the population use indicates how social media's role is seen: as the force that rubs the two opposing sticks of anger together, seeking the spark that can start the fire.

And it did work. An analysis for ANCIR suggested that the 'Guptas' fake news campaign' had ousted two finance ministers (one of whom had warned that billions of rands were being illegally shuttled to foreign bank accounts); the tax investigation service had been subverted to ignore corruption; and one of South Africa's biggest coal assets had been bought by a Swiss-based shell company – all helped by a news and social media campaign that made people focus on white capitalists, rather than corruption regardless of race.[44]

The other effect on democracy arises from external state interference via social media, where stirring up mistrust and division can bring foreign policy benefits through destabilisation of a country that either poses a threat, or where the foreign power can then masquerade as a friendly source of funding, trade or other support. The advantage of social media for external propaganda is that it's cheap to run and very difficult for the average person to detect: how can you tell at a glance who set up a Facebook Page boosting or booing a particular candidate, or whether that inflammatory tweet about a particular candidate even comes from someone who is eligible to vote?

Yet the social media companies have access to all the information they might need to detect such operations, such as the IP addresses from which posts are made, the frequency of posts, the interface used to post, the date and time and location of the account's original creation, and much more metadata that is invisible to the ordinary user. That enables them to act against 'inauthentic' accounts, which usually move together in packs, and target countries where familiarity with the internet, democracy or government – or all three – may be rickety.

'Africa is an enormous hotspot for these disinformation campaigns,' says Wilson. 'Especially the Central African Republic in 2019, where you can practically count the number of people on the internet on your two hands. It's so behind most of the world, and yet it was the focus of a concentrated Russian pseudo-state operation to push a whole bunch of propaganda into the country online. The technology allows external actors to amplify and influence in ways that with traditional media simply weren't possible.'

Social media is no longer a surprise to dictators or autocrats. They've adapted to the shock of the new and bent it to their will. 'What we've seen in the Philippines, Brazil and, frankly, the United States is social media in particular has been used as a tool of leverage for extreme populist parties in one form or another,' Wilson told me. 'It's allowing elites who are outside of the traditional political power structure to directly appeal to their own base, and build that independently, rather than working their way up politely through the party structure. They can come in as "an outsider", and generally speaking that narrative functions hand-in-hand with populism.'

Fred Turner, the Stanford professor of communications, knew that the hope of social media as the 'many-to-many' antidote to populism and fascism wasn't true, for he was writing soon after the election in the US of Donald Trump. 'Trump's capture of the presidency has visibly betrayed the anti-authoritarian promise of digital media,' he observed. Trump's social media presence effectively blocked out the sun for rivals, first to be the Republican candidate and then for the presidency, because his tweets generated so much media attention. It didn't matter whether what he said was true or (as almost always) false or bombastic; the extremity of the emotion enabled the media to discuss first the content, and then the fact that people were reacting to the content, and then whether the reactions of the people were correct or wrong, to fill the hungry hours of all-day TV. Mass media feeds off social media; the outrage that marks out social warming is often amplified for effect, rather than tamped down, by the mass media.

There's a well-trodden path down the road to populism, which has been used since long before social media existed. But, explained Wilson, social media makes it easier to spread – and the outrage mechanics and algorithmic amplification help it along.

'The notion of how democracies get co-opted by extremists is almost always through ethnic violence and threats,' he explains. 'In, say, Yugoslavia in the 1990s you had right-wing extremists who were insisting, in blatant propaganda, that the ethnic other is planning to take power, and when they do, they'll put all your children in camps, or round them up and ban them from even learning the language you speak at home. The detail doesn't matter – they just make up whatever they want to.

'The reason it works is, if you work it just right then the normal, moderate people who make up the bulk of society won't necessarily believe it. But they'll have a sort of weighted probability: look, they'll say, we're pretty sure the Bosnian Muslims aren't planning to do that. But *if* they are, the outcome would be *so* terrible that we've got no choice but to support the right-wing party, *just in case*. And the right-wing party just keeps pushing it further and further, because their incentive is to make the consequence of not backing them as incredibly outlandish and horrible as they possibly can.'

Forcing the opponent to deny they're going to do something terrible (so that they're then associated with the terrible thing) is an age-old political method. (In the US in the 1960s, Lyndon Johnson is famously said to have spread a rumour that his opponent fucked pigs.[45] His horrified campaign manager said 'Lyndon, you know he doesn't!' To which Johnson replied, 'I know – I just want to make him deny it.') What's different about them in the age of social media is that the lies get algorithmic spice sprinkled on them to get them to travel further, faster.

Are these problems inevitable, though? In June 2018, Katie Harbath, Facebook's global politics and government outreach director, spoke at an event in Berlin about protecting the integrity of elections.[46] 'I one hundred percent agree that the Philippines was Patient Zero,' Harbath said breezily, pulling up a slide about the five areas that the company now tried to focus on – foreign interference, fake accounts, advertising transparency, the spread of fake news and 'supporting an informed electorate'. It was a stunning admission of quite how wrong Facebook had been about its effects on democracy; all the more so because Harbath's interlocutor was Maria Ressa, the object of so much partisan hatred in her homeland. Harbath remarked that Facebook had about 10,000 people working on removing fake accounts, and that would double in 2019. But she also accepted that the elections team had its work cut out: the company operates in so many countries that there's always an election campaign going on somewhere.

In September that year, Ressa went to a high-level meeting at Facebook at its offices outside San Francisco, which was intended to discuss how social media content online can lead to harm in the real world. It happened just as the UN was publishing its report on the Myanmar genocide, in which Facebook was criticised for helping raise ethnic tensions.

According to a deeply sourced report by Julie Posetti, global research director for the International Center for Journalists, Ressa told the executives at the meeting: 'If you don't change what you're doing, I could go to jail.'[47]

'Facebook's mission is to make the world more connected and transparent,' as its cult-like mantra goes. But the second part is often missing. Speaking in July 2019 to CNN, her employer before she set up Rappler, Ressa commented, 'At least when you're in a war zone, the gunfire's coming from one side and you know how to protect yourself.'[48] By contrast, the information warfare being waged on her meant 'You don't even know where the enemy is here.' She warned that democratic freedoms were being eroded 'in front of our eyes': Rappler journalists had to attend Duterte rallies in disguise, because they weren't welcome; Ressa referred to the continual harassment online and from the government as being 'like pollution in the air'.

Less than two years after she had gone to the meeting at Facebook, Ressa was convicted on a charge of 'cyber libel'; the story over which she was convicted pre-dated the introduction of the law. Patient Zero was looking increasingly unwell.

Yet even all those examples probably understate the problem significantly. After ten years working for the CIA and the White House, Yaël Eisenstat joined Facebook in June 2018, with the job title of 'head of Global Elections Integrity Ops' – a buzzy phrase that she thought would mean she could help Facebook navigate the rapids of attempts to influence elections. Instead, she found herself effectively sidelined from her second day – and her first day had been the standard orientation that all new employees go through. The company was always reactive, always worried about offending those currently in office, and always more concerned that any solution be 'scalable' – a one-size-fits-all answer to any problem – than that it was appropriate to the specific country and situation where it was being applied. 'We could never address the unique intricacies of any particular country's elections or political issues,' she recalled in July

2020.[49] Eisenstat left the company after six months, frustrated and annoyed that she had not been given the power she had been promised: 'I was very clear as they interviewed me for the role that what I cared about was Facebook's effect on our democracy,' she said. But those above her never gave her the power to influence that.

Nor was she alone in her frustration at Facebook's indifference to its influence. In September 2020, Buzzfeed News published extracts from a memo written by a departing Facebook data scientist who had worked on detecting and removing fake accounts being used to influence elections, voters, democracies and governments.[50] Sophie Zhang said that in her three years at the company she had tackled influence operations in Azerbaijan, Bolivia, Brazil, Ecuador, Honduras, Spain, India, Ukraine and the US. They were often well funded, with disinformation and smears being generated on a full-time basis against political opponents by employees of political parties. Yet she also felt terrified of her own power to permit or deny such operations: 'most of the world outside the West was effectively the Wild West with myself the part-time dictator,' she wrote.

Zhang was hugely frustrated: Facebook as an organisation 'didn't care enough to stop' many of the influence operations, instead focusing on removing spam. 'The civic aspect [of Facebook's effects] was discounted because of its small volume, its disproportionate impact ignored,' Zhang wrote. Because to Facebook it's more important that people trust Facebook not to let them be scammed than that democracy be protected. Only the threat of having an incident cited by politicians or appearing in the national press would galvanise a coordinated internal response, she wrote, complaining that short-term decisions were 'largely motivated by PR and the potential for negative attention'. In effect, the security of democracy around the world depended on how likely a country was to be reported on in an American newspaper.

★ ★ ★

Facebook could stop this; YouTube could stop this. The machine learning systems could be tuned to spot posts about political candidates, or inflammatory comments about them, and delay them or introduce friction – a pop-up question, a warning – that might slow people, whether misguided or malicious, who were trying to take control of the discussion. It can already spot fake accounts, and prevent them from even setting up. Political discussion would be harder to patrol, but not insuperable. The failure is in distinguishing the coordinated from the authentic, and to realise that amplification algorithms are not appropriate for creating a level playing field.

'These are emerging democracies,' said Ressa, in an interview in September 2019. 'We don't have institutions, or if we do, they're extremely weak, which means we're far more vulnerable. That is part of the reason disinformation or information warfare in our part of the world leads to immediate violence.' Social media had become a danger, she said: 'if democracy is the body, they've allowed a virus in, and it's replicating quickly, and it is killing our body politic.'[51]

How can we tell whether people are going to use social media and their phones to organise in a 'good' way – to increase their access to democracy and have a voice in what happens, as happened in the Arab Spring protests in 2011 – or if they're going to use them in a 'bad' way, to back autocrats who will gradually erode their freedoms (and especially their enemies' freedoms)? John Stuart Mill never had to reckon with well-funded disinformation campaigns amplified by algorithms in a world where access to lies was free but checking for the truth was expensive.

Trying to tease out the precise effects of social media influence campaigns on election results, internal or external, probably misses the point, says Wilson.

'Everyone wants to know, did Russian interference on social media make it so that Trump won [in 2016],' he says. 'That would be the spy novel answer, wouldn't it. The bottom line is that the election turned on 50,000 votes in three swing states. Now, 50,000 votes were changed in three swing states based on the fact there were thunderstorms too. There's like twelve other factors that perhaps had equal influence, it's all within the realm of noise. But what we need to be focused on, the point of it, isn't whether we could ever conclusively prove that it threw the election to one side. The point is there is a foreign operation intentionally trying to affect democratic outcomes. It's not "Did Russia manage to swing an election?" The point is that they're trying. On a metaphorical level, on a political level, on a democratic level, this is an act of war.'

9

PANDEMIC: INOCULATED AGAINST TRUTH

We're not just fighting an epidemic; we're fighting an infodemic – Tedros Adhanom Ghebreyesus, director-general, World Health Organization, 13 February 2020[1]

It was 8a.m. in Toronto on a balmy June day, but Craig Silverman was already worried. To be fair, that was my fault: I'd asked him what his greatest concern for the coming year was in relation to social media.

'One of the things I'm very worried about right now is what happens when we get a vaccine for the coronavirus,' he told me. (We were speaking before any had been approved.) 'I think that the groundwork has been laid for years but is especially being laid now to really undercut people's confidence in any vaccine that emerges for it.'

As Buzzfeed News's media editor, Silverman has for years been fascinated by the ebb and flow of truth and falsehood over social networks and through news outlets. His first book, *Regret the Error*, in 2007 built on his website of the same name, which collated terrible mistakes newspapers made and then quietly corrected. The book's subtitle: 'How Media Mistakes Pollute the Press and Imperil Free Speech'.

But he thinks it's not the media that poses the big problems right now. Instead, the danger comes from intentional untruths spread on social media.

'That's where the anti-vaccine community is organising, and where it is acting like something of a gravitational force, pulling in people who might be unsure of vaccines, who might not understand the importance and the history of vaccines, to pull them in to make them "vaccine hesitant", "vaccine sceptical" and then, eventually for some, ardently anti-vaccine,' he said. 'And so I am worried that in a year, a year and a half, two years' time, the vaccine is available, it's produced in widespread quantities, but globally people don't want to take it because they've basically been radicalised by social media.'

Such a scenario might have felt almost inevitable if the social networks had stuck with their usual position: that they could not and should not decide what was true. In the first week of September 2018, Jack Dorsey, Twitter's chief executive, had told the US Congress that 'We believe it's dangerous to ask Twitter to regulate opinions or be the arbiter of truth.' In October 2019, Mark Zuckerberg told students at Georgetown University that 'I don't think most people want to live in a world where you can only post things that tech companies judge to be 100 percent true.'[2]

Yet within months of that speech, Facebook, YouTube, Twitter and other social media companies would execute a screeching handbrake turn on policy and be doing exactly that. Companies that had for years insisted they were neutral players, simply the blackboards on which people could chalk up differing opinions of all stripes, found themselves deciding about that most slippery of subjects, truth. When it came to the coronavirus, social media companies decided that they could, should and would determine what was true.

When the coronavirus first emerged in China, discussion on local social media was ruthlessly suppressed. But once the disease spread past the country's borders, the discussion moved to social networks indifferent to accuracy, and algorithmically enthusiastic about controversy and outrage.

On 27 January, four days after the city of Wuhan was put into a lockdown, the World Health Organization declared that the coronavirus risk was 'high at a global level'. The social networks noticed. Two days later, Twitter announced that searching for 'key terms' related to the topic would link to 'authoritative health sources'.[3] It also said that it would be watchful for attempts to spread 'platform manipulation' – what most people call misinformation and disinformation.

The next day, as the World Health Organization declared the coronavirus to be 'a public health emergency of international concern', the highest level of health alert, Facebook's head of health Kang-Xing Jin published a blogpost pointing out that, as before, content rated as false by outside fact-checkers would have its virality cut back – and, more significantly, announced 'an extension of our existing policies to remove content that could cause physical harm'.[4] Under this new interpretation, ideas that put you off getting treatment for the coronavirus, or which encouraged you to catch it, or which claimed to be cures (none were known) would be removed, not just downranked.

This was a remarkable departure for Facebook and for Instagram, to which the new rule would also apply. After years and years spent resisting or ignoring external demands to remove misleading content that could lead to harm, such as 'alternative' cancer treatments, autism 'therapies' and anti-vaccination screeds – where the children who would be harmed by believing them didn't even have the agency to determine whether or not to act – Facebook was now charging headlong into a content war.

The scale and the speed of response was notably at odds with that of many Western governments. It also demonstrated that the social networks were effectively dictatorships that could make unilateral decisions directly affecting billions of people worldwide without any recourse to public opinion.

Zuckerberg had been told early on about the risks the coronavirus posed because he and his wife, a paediatrician, fund an ambitious project called the Chan Zuckerberg Initiative that aims to make all disease 'manageable' by 2100. That project includes a number of eminent doctors, including Tom Frieden, a former head of the US Centers for Disease Control and Prevention. 'I was getting notes from them saying hey, this looks like it's not going to be containable,' Zuckerberg explained later in an interview.[5] 'It looks like this is really going to spread everywhere and is going to affect every country in some way.'

He told Facebook senior staff to be ready for how they worked to change dramatically, a warning that according to Nick Clegg, Facebook's chief of PR, was 'well ahead of what anybody within Facebook – and what folks other than the real expert epidemiologists – thought'. The moves on content were the most visible external change for the next month until the ramifications of the virus's spread became obvious to the rest of the world as well.

Nothing had changed functionally about how the social networks operated. But they did begin seeing themselves differently: 'Covid has seen us change our approach, and the phrasing is to "broaden our definition of harm",' one well-placed source in a social network told me. 'We see ourselves now more in a role of educating the public – I think during this Covid crisis, a lot of companies have actually taken on that more direct educational role.'

The World Health Organization set up a 'risk communication team' to distribute accurate information in the face of what they knew would be an onslaught of inaccuracy, misunderstanding and untruths. 'We know that even in the Middle Ages there was this phenomenon,' Sylvie Briand, who set up the WHO strategy, told *The Lancet* journal.[6] The question was whether humans had learned to trust experts in the centuries since the Black Death roamed the world.

Facebook already had experience dealing with an outbreak of untruth and misinformation about a dangerous viral disease – and, in the eyes of many scientists, had been found wanting. In May and June 2016, the Zika virus had spread north from Brazil, via infected mosquitoes, to Florida. Since that February the outbreak had spread through thirty countries, mostly in South America, with potentially serious effects for both adults and children. Facebook had done poorly in tackling it: 'misleading posts were far more popular than the posts dispersing accurate, relevant public health information,' reported a team led by Megha Sharma, a paediatric doctor based in Wisconsin, in November that year.[7] The differences were stark: the most-shared post with useful information, a WHO press briefing video, had been viewed 43,000 times, and shared by just under a thousand accounts. The most-shared misinformation post, which said Zika was a 'fraudulent medical hoax', had more than 530,000 views and had been shared by nearly 20,000 people – a twelvefold difference in views, a twentyfold difference in sharing. Years later, misinformation about Zika was still circulating in Brazil, to the frustration of doctors there.[8] That didn't augur well for how people might get their information about a disease caused by a brand-new virus with transmission mechanisms that were initially poorly understood.

The seeds of that problem were sown long ago, Silverman told me. 'I had a conversation with a person at Facebook in 2014 when I was doing research about viral rumours and hoaxes and debunking,' he said. 'We had a website where we showed our data, at emergent.info. The person at Facebook asked me: "So how are you identifying these rumours, how do you debunk them?" And I explained, "Well, you know, we're doing journalism, we're tracking. We use some Google AdWords, and we've built some Twitter lists, but at the end of the day, we do verification as journalists and researchers." And they were *very* disappointed. Because they wanted something that would scale, and that was automated. Facebook got so big because it cared about growth, and only growth, for so long. It's created its own trap where it's impossible for it to actually address the problems of misinformation and disinformation unless the solution could be scaled to many, many countries and work at a massive scale.'

The fact that this was a pandemic did therefore bring one benefit: Facebook was dealing with the same disease everywhere. The problem had already scaled to many, many countries.

Rather as the coronavirus had already slipped under the radar of those in the US – the first confirmed case was in Seattle on 21 January, before the announcements by Twitter or Facebook – so the problems that would become endemic to the social networks had already been seeded: conspiracy theories about the origins and effects of the virus, and about who stood to benefit from it. A pandemic is an epochal event, and the conspiracy theorists were ready to make the most of it. Social networks became a Darwinian breeding ground for the wildest ideas with the greatest resonance. How about: coronavirus was a bioweapon, produced in a Chinese laboratory not far from the Wuhan seafood market and unleashed on citizens there? Or: a bioweapon, but produced by the US military and unleashed on the Chinese citizenry? Or: the virus was introduced on purpose by the US 'deep state' in 2018, to be cured by a vaccine patented in 2015 by the Bill & Melinda Gates Foundation, which would profit hugely from curing the illness, but only when it had been used to oust Donald Trump. (You have to admire the far-sightedness such a plan implies.) Or: the coronavirus was energised or activated by the turning on of 5G mobile networks around Wuhan in November 2019. (In one variation, the activation of the 5G networks 'turned on' the virus while it was still in bats so it could spread to humans. No mechanism was offered for this process, apart from diagrams showing radio waves, bats and the word 'coronavirus'.)

The origins of this strange yet pervasive meme seems to have been a Belgian GP in a province of Antwerp, on 22 January. An investigation by James Temperton, writing for *Wired* magazine, found it in a (subsequently deleted) story in a regional version of a Belgian newspaper, *Het Laatste Nieuws*.[9] The local paper's usual fare was stories like 'Removing traffic lights should make intersection safer' and 'Interactive movement wall teaches children to count during gymnastics'. But here was a story about a GP in the district of Wilrijk with the headline '5G Is Life-Threatening and No One Knows'. The GP insisted that 'There is sufficient scientific evidence that the roll-out of 5G is damaging, but no one knows it.'[10] To anyone who has been writing about mobile phones for any length of time, the fears expressed were completely familiar: the 'radiation' involved was (somehow) dangerous, doctors were worried, we should err on the side of caution. The same line has been trotted out for at least twenty-five years, but no epidemiological link between mobile phone use and any illness or disease has appeared.

So, the reporter asked, might there be a link between 5G and the coronavirus, with the worldwide toll (by then) of nine dead and 440 infected, in Wuhan, a city with 5G towers around it since February the previous year? 'I didn't do a fact-check on it,' said the GP, vaguely. 'But it may be a link with current events.' The story was published in the print edition and online – and then deleted from the paper's site when the editor realised it had no scientific basis at all. But the damage was done. The genie had wriggled out of the bottle, though no doubt another one would have let it out had that one failed. For this was an idea that was aching to be born: a new technology introduced in a far-off land, and a new disease in that same far-off land.

Between the dangerous tweets and Facebook posts, and the conspiracy theories, the social networks found themselves fighting the systems they had created. The algorithms were built to reward virality, not accuracy; speed, not careful reflection; and were optimised to make passing things on effortless, not tricky.

The conspiracy spread from Dutch anti-5G groups to English-speaking ones, across Facebook and YouTube, and then began coming to the notice of more prominent influencers, driven – as Temperton put it – by 'engagement algorithms that were smart enough to spot a viral trend but dumb enough not to notice the idiocy of its content.'

From the battleground of memes, the '5G virus' turned out to have particular resonance. 5G was a new, faster mobile communications technology whose usefulness wasn't entirely clear to most people, and the masts that house mobile network transceivers have long been the object of public mistrust. A novel virus of uncertain origin and a novel technology of uncertain utility were a match made in heaven for conspiracy theorists, who could push their ideas to millions of people stuck at home, fearful about what was to come and seeking some way to understand it. While the coronavirus began to rage around the world, conspiracy theories did the same on social networks. The question was, would the response control either?

On the afternoon of 16 March, as the number of Covid-19 cases in the US passed 4,200 and total deaths reached 100, a joint statement from Facebook, Google and YouTube, Twitter, Microsoft and its subsidiary LinkedIn, and Reddit announced they would be 'jointly combating fraud and misinformation about the virus, [and] elevating authoritative content on our platforms'.[11] False rumours about lockdowns in the US were already running rampant, and four days earlier China had entered the conspiracy theory field, with a spokesman for its foreign ministry repeating in a tweet a wild suggestion from a Canadian conspiracy site that the coronavirus had actually originated in the US and been taken to Wuhan during the Military World Games in October 2019.[12] The companies seemed to have their work cut out.

In a blogpost on Monday 16 March, Twitter's legal chief and 'customer lead' explained that they would be 'reviewing the Twitter Rules' in the context of the disease outbreak.[13] Two days later, the post was updated with more specific advice: the reshaped policy would 'require people to remove tweets that include ... denial of established scientific facts about transmission during the incubation period or transmission guidance from global and local health authorities, such as 'Covid-19 does not infect children because we haven't seen any cases of children being sick.'

The same day, Mark Zuckerberg spoke on a conference call with journalists about what Facebook and the other companies were doing.[14] He insisted that the company was 'very focused on making sure that misinformation doesn't spread', and explained they were 'basically partnering together [with the other companies] to identify misinformation and things like that, that we need to make sure that we're addressing. The collaboration with other companies has generally gotten a lot stronger on fighting bad content over the last few years,' he said, citing earlier work on counterterrorism, where AI systems were first to identify more than 98 percent of the content that was subsequently taken down.

On coronavirus misinformation, though, the content would only be tagged, not removed, unless it posed an 'immediate risk of harm'.

Zuckerberg explained: 'There's one hoax going around that if you think you have this, drink bleach and that will cure it. And that's terrible. That's obviously going to lead to imminent harm.'

So why now was Facebook able to take action against misleading medical content, yet not before? And why couldn't it do that for political misinformation? Zuckerberg insisted that 'We've never allowed things that would lead to imminent physical risk . . . the standard here is you can't yell fire in a crowded theatre.' A politician who incited violence or posted 'dangerous' health misinformation would see that content removed, he said.

Equally important, Zuckerberg said, was that in this situation 'there are broadly trusted authorities . . . who people just across society would all agree can arbitrate which claims are conspiracy theories or hoaxes and what's trustworthy and what's not.' That also distinguished it from politics, he said. 'For health misinformation during a pandemic or outbreak like this, that's probably one of the most black-and-white situations you could expect.' Political speech, by contrast, 'is probably the most difficult in terms of how you arbitrate and govern that kind of speech'. The implication seemed to be that conspiracy theories, misinformation and disinformation by political rivals would be tamped down.

This turned out to be an optimistic reading of Zuckerberg's words.

Twitter was already discovering the challenges of policing content for truth. On 16 March it had deleted tweets by the controversial former Milwaukee sheriff David Clarke, whose response to the news of bars and restaurants being closed was to tell his 900,000 Twitter followers it was 'Time to RISE UP and push back. Bars and restaurants should defy the order. Let people decide if they want to go out.'[15] That, and similar ones, were removed on the basis that they violated Twitter's policies against 'encouraging self-harm'. A rule intended to stop teenage girls cutting themselves was suddenly being applied to the roulette wheel of unknown infectivity. Twitter also removed a tweet by John McAfee, whose life after founding a software antivirus company could be described as 'interesting', in which he claimed that 'Coronavirus cannot attack black people because it is a Chinese virus'.[16]

But life in the new age of truth determination wasn't always as easy as dealing with the tweets from Clarke and McAfee. On 19 March, Elon Musk, the CEO of the car company Tesla, who then had nearly 33 million followers, pointed to a piece of news from China in a tweet: 'No new domestic corona[virus] cases in China', he wrote.[17] Asked by another user if he believed that, he added 'Based on current trends, probably close to zero new cases in US too by end of April'.[18] Another asked whether that would be after store closures, 'social distancing', travel restrictions and mask-wearing. He responded: 'Kids are essentially immune, but elderly with existing conditions are vulnerable. Family gatherings with close contact between kids & grandparents probably most risky.'[19]

Except children weren't immune. Two days earlier, the *New York Times* had published an article pointing to a scientific study showing that nearly 6 percent developed serious symptoms, and that one fourteen-year-old had died.[20]

So did Musk's tweet meet Twitter's criteria for 'denial of established scientific facts'? Would it be deleted? Twitter pondered briefly and decided not. 'When reviewing the overall context and conclusion of the Tweet, it does not break our rules,' the company told The Verge, a technology news site, but declined to elucidate.[21]

So was there some room for scientific speculation? A few days later, The Federalist, a right-wing libertarian site, published an article by an unlicensed dermatologist suggesting a better strategy for tackling the Covid-19 pandemic was 'controlled voluntary infection', like the 'chickenpox parties' of the twentieth century, which intentionally aimed to infect children with the milder form of the disease. Now, Twitter acted, first putting up warnings to anyone who tried to click through to the article that the link was 'unsafe' (which made it sound as though it was harbouring malware), and soon afterwards suspending the site's account until the tweet linking to the story was removed.[22] In effect, the story couldn't be found through Twitter after that, because every link posted on Twitter is mediated by the company's own software, which lets it block links at the system level. (The principal purpose is to stop illegal content, spam, malware and pornography – usually the four main problems for any site – from spreading.)

The subtle difference in adjudicating 'truth' now became clear. Musk was pointing out relative risks, while Clarke and The Federalist had encouraged people to do things that might put them directly at risk. (Twitter's 'Safety' account had tweeted on 18 March that it would remove 'Tweets that include the following: Content that increases the chance that someone contracts or transmits the virus'.[23] Musk's actually pointed out the dangers of gatherings; Clarke's and The Federalist's ignored them.)

Twitter and Facebook carried on ramping up their aggressive posture on content. On 30 March, Twitter took the dramatic move of removing a tweet by a country's president – Nicolás Maduro of Venezuela, who suggested taking a 'brew' to 'eliminate the infectious genes' of the virus.[24] It followed up by removing two tweets by Jair Bolsonaro, Brazil's president, endorsing hydroxychloroquine, then an essentially untested treatment for Covid-19, and suggesting people end social distancing. Facebook also removed a Bolsonaro video as 'misinformation that could lead to physical harm'. Twitter then took down a tweet by Rudy Giuliani, one of President Trump's confidants, suggesting hydroxychloroquine had a '100%' success rate for treating the disease. A week later, it locked the account of the performance artists Diamond and Silk for suggesting that 'we must be out in the environment', and that 'Quarantining people inside of their houses for extended periods will make people sick!' Twitter said this was 'in violation of our

COVID-19 misinformation policy.' For Twitter, the deletions showed how far it had come from being 'the free speech wing of the free speech party'.

Facebook too began attaching warnings to stories more aggressively than before, pasting 'False' notices over egregiously wrong content that went viral. But that led to another problem. If Twitter and Facebook were now policing content, and removing that which was dangerous, should we assume that everything we saw was right? This was the 'implied truth' problem that the team at Yale, Harvard and Regina had discovered in 2019: labelling fake news as such (in their experiment, with a 'FALSE' warning much bigger than Facebook's low-key 'disputed' tag) meant that anything *without* that label would automatically be seen, by implication, as true.[25] In an atmosphere where tweets from high-profile accounts were being deleted, the potential for implied truth to attach to the millions of other tweets and posts on the topic was enormous.

The reality was that the combination of virality and algorithmic amplification would always let nonsense flower and spread in the fertile ground of social media and outrun the networks' ability to control it. In the first week of April, a survey by Pew Research found that more than a quarter of Americans believed the SARS-CoV-2 coronavirus was developed intentionally in a laboratory – a belief that was strongest among those aged under thirty, those who never went to university, Hispanics, and strongly conservative Republicans.[26] It was weakest among those with university educations, aged over sixty-five, or at the most liberal end of the Democratic voting group. There was no evidence to support the claim: all the scientific research suggested it crossed over naturally from bats as part of wildlife trafficking to Wuhan's live animal market.

While the social networks were putting out metaphorical fires, mobile carriers were putting out literal ones: some people set phone masts ablaze, believing they were 5G-capable (some were, some weren't) and hence, by their logic, a risk. By the end of the first week of April, twenty had been attacked in the UK.[27]

This was the effect of social networks, and their ability to find and concentrate the thinking – flawed or otherwise – of small and large groups. Before the amplification of mad schemes by algorithms, such ideas would have struggled to gain widespread traction. Social networks normalised such behaviour because whatever your view, you'd always be able to find some people, or even a lot of people, who would agree with you. That made it easy to not question whether the ideas really stood up to rational examination.

The platforms' new approach to truth meant that fact-checkers were in high demand. A study by the Reuters Institute at Oxford University found that between January and March, the demand for fact-checking rose tenfold, while noting that the volume of misinformation almost certainly grew much more.[28]

Most of the misinformation was just 'reconfiguration', the report found, taking existing information and twisting it; but a little more than a third was completely made up. Politicians and celebrities generated only about 20 percent of the claims but more than two-thirds of social media engagement, demonstrating the power of attention. However, the researchers noted that they couldn't see into the dark pools of closed groups and messaging, which they reckoned could be a significant source of 'bottom-up misinformation'.

The worst at removing false content turned out to be Twitter, where 59 percent of posts rated as false were still up, against 27 percent on YouTube and 24 percent on Facebook.

The report's authors weren't hopeful: 'there will be no silver bullet or inoculation – no "cure" for misinformation about the new coronavirus,' they observed. Instead, there would have to be 'sustained and coordinated effort' by fact-checkers, media, public authorities and, notably, the platforms themselves. The eternal problem, though, was that fact-checkers could only move at a certain speed, while the sharing mechanisms built into the networks enabled information, whether true or false, to move much more quickly. Given that untruths spread faster and wider than the truth, fact-checkers were always going to be dealing with an exponentially growing number of claims.

WhatsApp took its own measures to try to restrict that early in April, as most of the world's population was becoming better acquainted with the idea of viral spread, by significantly reducing the potential for inflammatory messages to do the same. New limits meant that messages that had already been forwarded five times could now only be forwarded to a single new chat (which might be to an individual or a group) at a time.[29] Strangely, the company chose to headline the blogpost about this 'Keeping WhatsApp Personal and Private', when the intention was clearly 'Keeping WhatsApp From Being Used To Spread Misinformation'; the blogpost essentially admitted as much, noting that 'we've seen a significant increase in the amount of forwarding which users have told us . . . can contribute to the spread of misinformation.' That forwarding is a major conduit for misinformation was not news to WhatsApp, which had seen how that panned out in India and Brazil, to sometimes deadly effect. Reducing virality – not a phrase the blogpost used – could 'keep WhatsApp a place for personal conversation'. Retreating from virality was all the rage. At the end of the month, the company said there had been 'a 70 percent reduction in the number of highly forwarded messages sent on WhatsApp'.[30] Such a drastic fall implied that the most viral messages had previously been passed on dozens of times. And from what we already know about the nature of information cascades on social networks, those were twice as likely to be false as true.

Frequently, the tech companies' focus on information about coronavirus only served to highlight the damage caused by their laissez-faire approach elsewhere. On 10 April, putting a search for 'coronavirus' into Google yielded just a page of straight, factual information – government advice, WHO advice, fact-boxes about prevention. The suggested searches were topics like 'coronavirus UK', 'coronavirus US', 'coronavirus stats'. Even a search for 'coronavirus 5G' turned up no conspiracy theories on the first page, only debunkings. A search for 'herbal corona cure' produced a fact-box at the top of the page that said, 'There is no specific medicine to prevent or treat coronavirus disease (Covid-19).' There were more fact-boxes with details about self-care and medical treatments, the latter boiling down to 'get in touch with a doctor'.

People noticed, and approved. But they also noticed that Google wasn't always doing the same against a search for 'herbal cure for cancer stage 4' – which would produce millions of results, the top four being lucrative adverts, including 'Baking Soda Cancer Treatments' and 'How to cure stage 4 cancer naturally'. (Stage 4 cancer is usually the prelude to death within a matter of years or months.) Dan Olson, a screenwriter, said he had started typing 'herbal cure' into Google, which added 'cancer stage 4' as one of its top five suggested searches.

'I just want to point out that Google could do this [diverting searches to authoritative sources] for all quack medicine, but they don't,' he observed on Twitter.[31] The adverts pointed to a reason why it might not: why bother being authoritative when you could be wealthy?

Facebook had precisely the same problem. On 16 April, Zuckerberg wrote a blogpost celebrating how many warnings had been displayed on posts relating to Covid-19, and saying that 'Through this crisis, one of my top priorities is making sure that you see accurate and authoritative information across all of our apps.'[32]

On the same day, Aaron Sankin was browsing Facebook when he was shown an advert: 'Keep your noggin cozy and clear of cell phone radiation with a Lambs anti-EMF beanie!' it suggested, showing an ordinary-looking beanie hat that was, apparently, 'radiation-blocking'.[33] Intrigued, Sankin checked why Facebook was showing him the ad: the vendor was trying to reach people aged twenty-five to fifty-four living in the US who 'Facebook thinks are interested in pseudoscience'.

Sankin isn't really interested in pseudoscience, except by accident: he's an investigative journalist at The Markup, and had begun looking into the plethora of coronavirus conspiracy theory Groups that had been set up on Facebook. But the advert revealed that not only had Facebook created its own 'pseudoscience' category for ad targeting, but that there were as many as seventy-eight million people who could be reached through it in the US alone. An ad he created targeting them was approved in a few minutes; the same for Instagram. What surprised Sankin more was that the advertiser for the 'anti-EMF beanie' told him they weren't aware of the category; Facebook's algorithms had slotted it into that category as a way of reaching the best audience to fit with what it was selling.

In other words, the advertising algorithms were undermining the efforts of the humans trying to push rational, scientific approaches to the world.

The day after The Markup contacted Facebook to ask about the pseudoscience advertising category, the company removed it 'to prevent potential abuse in ads'. How much might those ads have been worth? The Markup didn't make an estimate, but based on Facebook's advertising revenue in the US and Canada of $34.1 billion in 2019, those seventy-eight million people, out of 250 million users, could have been the target of $10 billion in advertising in their own right.[34] For Facebook, they were clearly a desirable demographic.

Left unexplained was why the company had created the category if it thought there was potential for abuse – and why it had been allowed to exist since 2016, as data collected at that time by the news site ProPublica showed. (Also available at the time were advertising target categories such as 'New World Order' and 'chemtrail conspiracy theory', though they were subsequently removed.) In 2019, the *Guardian* had found that advertisers could target people about 'vaccine controversies'. Facebook's commitment to scientific accuracy had always looked loose at best; now it looked more like a place happy to encourage conspiracy theories, or at least monetise them, no matter what public pronouncements its leaders made.

The trouble with encouraging conspiracy theories, though, is that once established, they're impervious to facts, and especially rebuttal: 'Rejection by authorities is [for believers] a sign that a belief must be true,' as Michael Barkun's seminal 2016 paper on the topic pointed out.[35]

There was another reason to worry about conspiracy theories, or more specifically the people who believe them. A study by King's College London in April 2020 found that such people tended to be the ones who weren't inclined to follow scientific and government advice to wash their hands well, stay at home, and use 'social distancing' to minimise the risk of spreading the virus. It found that 5 percent of adults in a random (though not weighted; two-thirds were women) poll believed that 'the symptoms of Covid-19 seem to be connected to 5G mobile network radiation' (plus nearly 10 percent who thought the pandemic was planned by pharmaceutical companies and government agencies, and 24 percent who thought the virus 'was probably created in a laboratory').[36] Those who believed the conspiracy theories, particularly the 5G idea, were notably less likely to trust the medical advice – with more than one-third of the 5G believers thinking there was no good reason to stay at home. The larger group who thought the pandemic was 'planned' actually posed the bigger risk to public safety: a quarter of them, or 6 percent of the population, were in the 'no good reason' camp.

With that in mind, posts and videos that push conspiracy theories suddenly look less like the joys of free speech, and more like a risk to people's health – an excessive form of social warming. Belief in conspiracy theories also tends to snowball: believing one makes it easier to believe another. After all, if you think there's a huge conspiracy by evil overlords to cover up one thing, then why wouldn't they also cover up or control two things, or three or four? In that context, the recommendation algorithms at work on the platforms begin to look like an engine to make some people believe things that will actively make them ill.

But at least you can sell advertising space encouraging them to buy beanie hats along the way.

What conspiracy theorists might have found more shocking, away from the irradiated bats, was the extent to which the social networks were coordinating their initiatives so that new measures would be taken within days or even hours of each other on the respective platforms. Within a few days of each other early in March, Google and Facebook and Instagram began banning adverts for medical face masks, to try to stop profiteering.[37]

'There's a weekly call and we share trends,' one well-placed insider at a platform told me. 'But we don't coordinate down to the level of "You're removing X, so we'll remove X." We all have different rules, and you'll see press coverage that says, "Platform A took it down, Platform B put it behind a [fact-checking] wall, Platform C did nothing" – that happens quite regularly.'

They were careful not to call it coordination; the heightened attention being paid to them meant any hint that they were agreeing exactly on what could and could not be on their services would have provoked demands for more congressional hearings and investigations.

The war between the social networks and the purveyors of disinformation intensified in line with the epidemic. On 1 May, Facebook removed the Page of David Icke, a conspiracy theorist and former professional goalkeeper, for 'repeatedly violating our policies on harmful misinformation'. Which of Icke's posts and videos had tipped the balance wasn't immediately clear: perhaps the suggestion that 5G mobile networks were spreading the coronavirus, or the video suggesting the virus couldn't be spread by shaking hands. The next day, YouTube deleted Icke's channel, also for repeatedly breaking the rules on 'content that disputes the existence and transmission of Covid-19 as described by the WHO and NHS', a spokeswoman told the BBC.[38]

But it was a conspiracy theory version of whack-a-mole. A few days later, a video began taking off on social media featuring a discredited former research scientist, Judy Mikovits, claiming that the pandemic was planned – a 'Plandemic'. The thirty-minute video was linked or hosted on Facebook, YouTube, Instagram, Twitter and LinkedIn. Besides claiming that wearing face masks could make you sick and that beach sand conferred coronavirus immunity, it also suggested that any vaccine against the virus would itself be dangerous.

The video and associated book saw rocketing views and sales – and immediate action by all the social networks, which removed copies from their sites. 'Suggesting that wearing a mask can make you sick could lead to imminent harm, so we're removing the video,' a Facebook spokesperson told Buzzfeed News.[39]

One group was quick to take advantage of the follow-on from the eruption of infections and deaths, and the hopes of most of the population that a vaccine could be developed that would allow life to return to normal, or at least its simulacrum. The anti-vaccination fringe saw the widespread hopes of a vaccine that almost all of the population would require as indicative of a deeper motive: a conspiracy. Remarks plucked from speeches by Bill Gates and the US health director Anthony Fauci, who had both warned in the previous decade that the elapsed time since the 1918 flu pandemic plus the growing interconnectedness of the world meant humanity was overdue another one, were used to weave a narrative in which both men were suspiciously prepared for an outbreak. Gates's humanitarian work of funding extensive vaccination efforts in developing countries, and thus saving millions of lives, was depicted as an evil plot to control the world by making vaccination mandatory, and including a microchip 'to track people' with each shot.[40] (Humorists pointed out that the microchip wasn't necessary, as everyone carried around smartphones that did the job already.)

While sensible adults rolled their eyes and pointed out that the pandemic-stricken present was what a world without a vaccine looked like, the anti-vaccination groups connected their wild ideas in what they saw as a cast-iron proof of intent and implementation. The fact that the social networks kept deleting their content was only more evidence of the conspiracy, because to believers no conspiracy theory is ever wrong – just incomplete.

Buzzfeed noted in May 2020 that the increased activity by anti-vaccination agitators looked to the algorithms on Facebook and Instagram like the higher engagement their systems are attuned to, especially if people who were pro-vaccination began interacting with the antis on their Pages or accounts.[41] Algorithms struggle to discern the difference between furious disagreement and delighted approval, especially if both have a picture of a baby.

That inability to distinguish encouragement from opposition had already infuriated some mothers, who had uncovered a seething mass of exploitation around childhood vaccination and autism, which are often falsely linked. In 2013, Melissa Eaton began wondering about her son, who was approaching his fourth birthday: his behaviour was subtly different from that of his peers. 'Doctors were mentioning "sensory processing disorder", and I sought information and support online,' Eaton, who lives in North Carolina, told me. On Facebook, she found that questions posed in parent-led support groups would get responses trying to pull people over to 'biomedical' support groups. 'Someone suggested a cilantro [coriander] and frankincense enema for my son, and wanted to tell me how to do it – for fifty dollars,' she told me. 'I began looking for better places for information, because I could tell they had no understanding and were offering harmful advice.'

That was just the beginning. In 2014, her son was formally diagnosed as autistic, and Eaton discovered that there was an entire industry hidden away on social media selling unproven or potentially harmful autism 'treatments' for credulous parents.

Since August 2018 – though at no point before – Facebook's rules have banned content that 'promotes, encourages, coordinates or provides instructions for use or making of non-medical drugs'. However, banning is one thing; enforcing quite another.[42]

'It takes one parent to be convinced and they spread the info to other parents in support groups with a "try this" approach,' Eaton explains. 'They invite parents into private Groups run by the sellers and these memberships grow to thousands. These Groups kick anyone out who dares challenge their "protocols and treatments" so it just becomes this huge echo chamber of misinformation. Once parents are in these Groups, it's hard to pull them back out.'

Autism is a neurological disorder with no clear cause, and no known medical treatment. But because a diagnosis can have devastating implications for a child's future, parents naturally search for solutions. And people on the Groups were happy to offer them: that the cause was actually parasites, or (of course) vaccines. A common suggestion for treatment was ingestion of 'Miracle

Mineral Supplement', or MMS, given orally or as enemas – an idea pushed in a book titled *Healing the Symptoms Known as Autism*, which was published in 2013 by a real estate agent based in Florida.[43] MMS is in fact chlorine dioxide, an industrial bleach, which can cause vomiting, stomach pains and serious blood disorders in people with certain enzyme deficiencies.

Some of the parents would share pictures taken after the enemas of what they believed were parasites flushed from their child's intestines. In fact, it was part of the intestinal lining, stripped off by the chemical. Eaton was so outraged that she teamed up with Amanda Siegler, a mother in Florida, to infiltrate the Groups, take pictures and report the parents to local child protection authorities.[44]

How well had the social networks responded? 'Facebook has been the worst,' Eaton told me. 'Their strong stance on not removing content exploded this environment. It's so unmanageable there now that it would take years to undo the harm if they acted now.'

By contrast, Pinterest had taken the strongest position, she said, filtering out results on the best-known harmful searches such as 'anti-vaxx' and the autism book author's name. 'They limit the search results and advise people to contact a health provider.' Twitter's lack of private groups was a double-edged sword: 'It's out in the open so everyone becomes a target for misinformation.' But, equally, everyone could see what was being done, and challenge it. 'YouTube since 2019 banned a lot of content using terms like MMS bleach, but fails badly at removing violations and banning repeat offenders.'

The biggest frustration Eaton faced was in getting a useful response. 'We can only get a social network to act if we get a high-profile reporter to bring it to them and ask for a statement,' she said. 'They then sometimes remove things to avoid bad exposure in the media. Oftentimes they only remove the page and not the personal profile, so the bad actors are able to warn some of their followers which new platform they are moving to.'

Anne Borden King, who has also worked with Eaton to tackle those inveigling people into buying pseudoscience 'remedies', expressed similar frustration. She was angered by the consistent stonewalling, as she sees it, of Facebook. 'Again and again, we'd get back "This doesn't violate our Community Standards". At some point you go, "Well, maybe their Community Standards is a PR game."'

An analysis by the lobbying group Center for Countering Digital Hate (CCDH) in 2020 suggested that Facebook has been used as a 'shopfront' for anti-vaccination campaigners, who also used YouTube conspiracy theorists to push their ideas.[45] 'Social platforms chose not to alienate an anti-vaxx user base that we estimate is worth up to $1 billion a year to them,' suggested Imran Ahmed, chief executive of the CCDH, in the analysis. Despite taking a number of measures against those publicly pushing anti-vaccination content, Facebook was still accepting paid anti-vaccination adverts and YouTube allowed people pushing those ideas to monetise their videos with adverts, the investigation found. (In October 2020, Facebook announced it would ban adverts discouraging vaccinations, though not advocacy about government policies on them; YouTube said it would remove videos spreading disinformation about coronavirus vaccines.)[46] The CCDH, like Eaton and King, found that many of the Facebook Groups pushing pseudoscience with a price tag were hidden, so that they could only be joined by invitation. Once inside, anyone who expressed contrary views would be expelled – creating echo chambers that could effectively radicalise people around the topic.

The harm isn't just theoretical. The risks of measles include death. And in the US, between 2014 and 2019, the American Association of Poison Control Centers recorded more than 16,000 cases of chlorine dioxide poisoning, including 2,500 in children under twelve. (How many are due to 'medication' can't be known:

hospital staff didn't record whether the children drank the bleach by accident – a common household accident – or were given it.) One included a six-year old girl with autism who in 2017 was admitted to hospital with liver failure – one of the chemical's potential side effects.

A study published in *Nature* in May 2020, led by Neil Johnson at George Washington University, looked at the organisation of anti-vaccination advocates on the site.[47] The researchers found around 85 million accounts involved around the topic of 'vaccination', of which nearly all were classed as 'undecided'; 6.9 million were members of pro-vaccination Pages, and 4.2 million of anti-vaccination Pages. But more Pages were anti than pro, and the number was growing faster; and, counterintuitively, an outbreak of a disease such as measles prompted more to join the anti-vaccination Pages. The back and forth around the 'undecided' participants was 'akin to a battle for the "hearts and minds" of individuals in insurgent warfare', the authors noted.

Yet the reality is slightly different, because most people get themselves and their children vaccinated. The US Centers for Disease Control and Prevention's statistics say that over 90 percent of children in the US have had polio, MMR, hepatitis B and chickenpox vaccinations, and over 80 percent have had Hib, PCV and the DTaP vaccines.[48] Childhood vaccination is overwhelmingly supported.

But just as Silverman feared, the rush in the US by the Trump administration to have a vaccine approved before the November 2020 election fuelled legitimate concerns about corner-cutting on safety: proposed vaccines have to be tested on huge numbers and a wide variety of people to guard against unexpected side effects. Anti-vaccination groups leapt on the reasonable worries and used them as recruitment fodder.

Yet by June, the pandemic did seem to have brought about a new realisation at the top of social media companies: that their approaches to truth and untruth could make a difference, and that that could be measured in lives. What if that could also be applied to content that tried to mislead, or to incite hatred? The middle of 2020 marked the end of the first stage of the pandemic, and became a moment when the social networks assessed the world they'd helped create and realised that there were improvements to be made in their own approach.

Among those pressing them was the Center for Humane Technology, a pressure group that tries to get tech companies to realign themselves with the public interest. Its members and advisers include former Googlers (including Guillaume Chaslot, the disaffected YouTube engineer), the co-founders of Pinterest and Lyft, and one of the inventors of Apple's Siri.[49] That makes it well connected to the thinking within these organisations as it tries to shift them. David Jay, the head of mobilisation, thinks that the pandemic gave the tech platforms the ideal opportunity to change their behaviour so they can stop being punching bags for politicians and the press.

'Facebook, especially since [the elections in] 2016, is eager to find a way to rebuild trust. I think they're hoping their coronavirus response can be a way to do that,' he told me. 'When I track the conversations happening inside of tech platforms, this is kind of a biblical flood.' In essence, he saw the pandemic and our consequent reliance on tech companies and social networks as giving them the opportunity – if they choose to take it – to rethink their reason for existing.

'I could see it being rewritten to say, "We used to try to be these neutral platforms, but now we have this clear sense of our social purpose." They could be saying, "We tried not to have an opinion but then it was co-opted by rising authoritarian governments." The mythos of why tech platforms exist is going to be rewritten because of this event for decades to come.'

Silverman, meanwhile, worries that the anti-vaccine community will make a difference that will be felt more immediately.

'There really is a powerful and very active anti-vaccine community that is organised and connected around the world,' he told me. 'With the pandemic, there's become kind of an unholy alliance between anti-vaxxers and the anti-pharma[ceutical], anti-science communities and conspiracy communities – whether it's the QAnon people or anti-government extremists – and all of them have found common cause in believing that the pandemic is not just an actual novel virus, but is a *planned* event meant to execute the subjugation of people around the world by corrupt elites. They all, in their own ways, believe that's true.'

This for him illustrates the two-edged sword of social networks. 'All of the wonderful power of social media bringing people together, of elevating voices, of enabling you to reach and target people in a more efficient way than ever before – that can be used for good, benign activity. But because it is fundamentally open in a lot of ways, and because these platforms are too big to actually effectively moderate themselves, it creates a massive opportunity for bad actors, and for manipulation.'

That's why he worries about anti-vaccine groups alighting on any bad news around a vaccine and magnifying it to suggest the entire industry is dangerous.

But isn't there another possibility: that people who face the choice of getting vaccinated and being able to return to 'normal' life, or refuse and be excluded, will give in to their better judgement, and tear down the anti-vaccine conspiracy?

'That would be a great scenario,' Silverman says. 'It's true, this could be an opportunity for a global reinvestment and reinforcement of the value to health and the safety of vaccines that have been well tested.' He pauses. 'But my worry is that the anti-vaccine community is treating this as their moment to seize. They have been working on this for months and months and months.' He cited the *Nature* study that showed how anti-vaccination Groups had been more effective at expanding their membership on

Facebook than pro-vaccination Groups. 'I think the anti-vaccine groups are being very strategic. They are fundamentally attacking the credibility of the institutions, the people, the foundations and the scientific community. So even if all those folks get their act together, they've laid the groundwork for conspiracies and other things to create doubt.' His concern was borne out in December 2020 when Margaret Keenan, a ninety-year-old woman in the UK, became the first person in the world to be vaccinated against coronavirus; on Facebook, some people immediately insisted that she had in fact died in 2008 and that the person shown was an actor.[50]

As ever, Facebook wasn't going to get in the way of people who wanted to create divisions in society. Interviewed by Axios in September 2020, Mark Zuckerberg was confronted directly with the question: given how quickly he had moved against misinformation on coronavirus, would he do the same and remove 'anti-vaxx' propaganda and misinformation that was sure to follow the announcement of any vaccine against Covid-19?[51]

'All the challenging questions about antivaxx, and frankly a lot of misinformation, come from cases where there has been some instance of harm, but that people are kind of blowing it out of proportion or saying it's more prevalent or common than it actually is,' Zuckerberg replied. 'So, is it true that sometimes vaccines get mixed incorrectly by a doctor and that causes harm? Or is it true that even when a vaccine is helpful for addressing something overall, that it may not always be 100 percent effective? Yes, these two things are true statements, but does that imply that you shouldn't go get a vaccine? In general, not.'

It didn't seem to be a ringing endorsement of the benefits of a better-informed public over Facebook's desire to let people post pretty much anything. And it wasn't.

'I think that if someone is pointing out a case where a vaccine caused harm, or that they're worried about it – you know, that's a difficult thing to say from my perspective, that you shouldn't be allowed to express at all. But what we try to say is if people are overgeneralising things, if they're spreading misinformation and we can defer to a third party fact-checker or organisation like the CDC or WHO on clear health guidelines on things that could lead to imminent harm if people don't follow them, then we will try to take action against those.'

As usual, he had both sidestepped the question – which had been about anti-vaccination propaganda on the grand scale, not individuals' experiences – and handed off responsibility, and set a high bar for removal. It would require a fact-checker suggesting that the content could lead to imminent harm before Facebook would consider acting; and that might only be 'downranking', not removal. And 'imminent harm' is impossible to prove in the case of a vaccine. Silverman's hopes would have to play out against Zuckerberg's implacable principles and the bubbling, insistent advocacy of the anti-vaccination groups. The pandemic had confronted the social networks with the enormity of their social responsibility in being able to transmit information to billions of people around the world in an eyeblink. Yet on the evidence, with conspiracy theories blooming (even after Facebook's crackdown on militias and QAnon) and disagreement about what needed to be done becoming more and more entrenched, it was hard to say that they had managed to prevent the natural tendency of their own systems to run out of their control.

10

REGULATION: CUTTING THE PROBLEM DOWN TO SIZE

The question has always been whether we'll be shutting down an industry unnecessarily, [yet] not whether we are risking unalterable changes in the atmosphere – Sherry Rowland, one of the scientists who discovered how CFCs damage the ozone layer[1]

Nobody meant for this to happen. The purpose was to help us find friends, to stay in touch with our families, to cultivate new interests, make the world more 'open' and 'connected'. Not to destabilise democracies, to incite the killing of innocent people and the displacement of entire populations, or to polarise societies and undermine politics.

But it is happening. The social temperature is rising. Disagreements and collisions of views happen more easily, misinformation and disinformation spread more quickly, outrage is weaponised, the people who are meant to be leading countries are instead being distracted by how people on social media, rather than the population at large, will respond to their policies. We are drawn back to the social platforms by a simmering unease that we are missing out on something important or interesting. Then when we start scrolling, we find ourselves overcome by outrage, polarised by artfully evolved scissor statements.

Even where the penetration of social media isn't widespread, its effects are significant. A comparatively small but vocal group can start or influence social trends as well as distorting national politics.

'People developed planes first and then took care of flight safety,' Mark Zuckerberg said in an interview in 2016. 'If people were focused on safety first, no one would ever have built a plane.'[2] He was talking in the context of artificial intelligence, and any risks it might pose, but the metaphor works for the attitude that social network executives have taken to their responsibilities: let's first take flight and then figure out how not to crash.

That approach, however, has brought measurable costs to society that everyone else has had to bear, while Facebook, Twitter, Google (through YouTube) and other social networks have grown rich. The warming effect of the networks – the by-product of their widespread, uncontrolled use exactly in accordance with the instructions – is observable everywhere. The effects on democracy alone, particularly the opportunities that they allow for foreign governments to influence election outcomes, warrant oversight in their own right.

Every day, around 1.8 billion people log into Facebook. In a month, if you add in users of Instagram and WhatsApp, the total is more than three billion unique users.[3] Twitter reaches 190 million users worldwide each day.[4] On YouTube, around five billion videos are watched each day, and during a month, more than two billion people visit the site.

These are numbers our minds struggle to comprehend. In simpler terms, in the US, about three-quarters of adults look at Facebook at least once a day, almost the same number look at YouTube, and one in five use Twitter.

Yet despite overseeing that colossal amount of interaction, the platforms are comparatively tiny employers. Facebook employs fewer than 60,000 people; Twitter, a little over 5,000. Google has about 200,000 staff, though it has never released figures for YouTube (since the two companies are not distinct). Even Google's numbers don't put it in the top thirty largest employers in the US. All these companies were created to grow fast and lean. In the hacker culture espoused in Silicon Valley, the most important question is 'will it scale?' – meaning, will we be able to offer this service to a far larger number of people without increasing our staffing or costs at the same rate? Social networks have been marked by the pursuit of growth at all costs, and ignoring the costs of externalities that can be denied.

The other Silicon Valley mantra is that 'it's better to ask forgiveness than seek permission,' denoting the headlong rush to try to change the world and only later work out whether that change is beneficial. If the negative changes are hard to attribute directly to the network, all the better: less forgiveness to seek.

But there comes a tipping point, when a social network scales beyond control. While the number of users grows arithmetically, the number of potential social connections grows geometrically; Metcalfe's Law, oft-quoted in cybernetics, suggests that the effect of a network grows by the square of the number of users. Put 1,000 people on a network, and you have one million potential connections. Add another 1,000 users, and now you have 2,000 users and four million potential connections. Your network is twice as big, but your moderation problem is four times larger. Add in the power law, and your most popular users (who might also be the source of your biggest problems) will have dramatically larger audiences, and the potential to cause even more headaches because any moderating action you take against them will be even more visible, and surely unpopular.

Even without an algorithm churning away trying to put people with shared interests together – ignorant as to their motivations – any attempt at moderating content will quickly be overwhelmed unless the system for doing it scales in line with those potential connections. If it doesn't, the network can't be properly controlled.

Facebook regularly tries to deny or deflect this suggestion. In mid-2020, as a boycott by large advertisers was gaining media attention, the public relations man for Zuckerberg's creation, Nick Clegg, insisted that 'the vast majority of those billions of conversations [on Facebook] are positive.'[5] That earned a sharp riposte from Julia Carrie Wong, a US-based writer for the *Guardian*, who has repeatedly pointed out the existence of Nazi and far-right networks on Facebook to the company itself (and been harassed by members of those networks as a result). How many Likes on a cute picture of a grandchild, she asked, would balance out the murder of a security officer by the men who met and planned their attack on a Facebook Group?[6] How many families connecting over Facebook during pandemic lockdowns would make up for the 175-entry-long list of white supremacist Groups she'd sent to Facebook, which evidently couldn't root them out itself? 'Over and over again reporters, researchers and activists have documented the real and devastating costs of Facebook's algorithmic negligence and record of accommodating hate,' she wrote. Saying that most conversations on Facebook are positive is like saying that three of the four nuclear power reactors at Chernobyl didn't blow up: it's true, but ignores what's actually important.

Facebook, and YouTube, and Twitter will not and cannot admit that they have lost control of their creations. The role that Julia Carrie Wong and so many other journalists and researchers have played in highlighting the networks' failures falls to them because there was never really a point when the social platforms were fully on top of their users' activity. That wasn't their intention; the aim was to get as big as possible, as quickly as possible.

One topic that came up repeatedly when I asked people how they thought social networks could – or should – be refashioned to make them less attritional was to make content sharing a little less easy. Chris Wetherell, who invented the retweet function at

Twitter, told me that malicious use seemed to be growing, and that concerns him.

At Cambridge University, John Naughton suggests that the networks need to make it harder, rather than easier, to spread content. 'The retweet button is one of the most destructive things there is. There are all kinds of things you can do which would slow it down,' he told me. 'If you're a control engineer then one of the things you know is that if a feedback loop gets very fast, then you can kiss goodbye to any idea of control, and we have clearly long gone beyond that point.'

There could be no better illustration of how Facebook has lost control than the emergence and spread of the QAnon conspiracy theory. This pushes the idea that children were being sacrificed for their blood by a powerful secret cabal, which in turn was being tracked by an even more secret group led by Donald Trump, who was always imminently about to announce a wave of arrests. (In essence, it was a reworking of an old anti-Semitic blood libel, with a sprinkling of Trump worship.) Having started on other message boards in 2016, the concept metastasised on Facebook. The FBI identified it as a domestic threat in May 2019.[7] Yet on Facebook the runaway train continued for more than a year; it wasn't until August 2020 that the company made its first concerted effort to remove and ban adverts, Groups and Pages related to the topic. The effort was renewed in October, this time going after accounts claiming to 'represent' the mysterious cause. In a statement, the company said it regarded QAnon as a 'militarised social movement'. Did nobody at Facebook, after years of being endlessly criticised for failing to act responsibly when enabling terrorism, feel that they should act earlier? 'Our timing was really based on the work we do to understand and combat violent threats,' a spokeswoman told one journalist who asked that very question.[8] In other words, Facebook is too unwieldy to respond to threats identified by a domestic intelligence agency. It's too unwieldy, indeed, to respond to the continual problems that its externally employed moderation team of around 15,000 flag to it: 'Pretty much the only language Facebook understands is public embarrassment,' one moderator told the *New Yorker* in October 2020.[9]

Twitter and YouTube, and Facebook's other properties Instagram and WhatsApp, aren't innocent either. Twitter's incipient tendency to allow harassment of women and minorities was a side effect of its creation by a workforce that was largely white and male, who ironically tended to be less intensive users of their own service – and so even less likely to experience its problems – because they were so busy building it. (As an example of how blind its staff could be to potential problems, near the end of 2018 they suggested having a 'presence indicator', which would show that someone was online; it harked back to the service's origins as a tweak of early-2000s instant messaging.[10] Public reaction to the idea was almost unanimously negative: stalkers and harassers, they pointed out, would have a field day because they could track their target's activity. Twitter decided not to implement it.) YouTube's endemic problems are well documented. Again and again, though, the companies shrug in the face of the sheer scale of the problem that they have created for themselves.

There's also a reciprocal reluctance on the part of the platforms to tangle with the powerful. As we've seen, Facebook and Twitter have very publicly declined to censor content published by political leaders that, if it came from an ordinary user, would get them banned. In June 2020, one Twitter account put this directly to the test, tweeting the same words as Donald Trump. It was suspended within three days for violating the rule against glorifying violence, while Trump's account remained untouched.[11]

Treating voters differently from those they elect is shamelessly anti-democratic. Twitter and Facebook have both justified leaving such leaders' content on their platforms on the basis of its 'newsworthiness'. However, as a group of former Facebook staff pointed out in a letter to the *New York Times* in June 2020, if what politicians say is by definition newsworthy, and newsworthiness means content must be left alone, then 'there is no line the most powerful people in the world cannot cross on the largest platform in the world.'[12] Such special treatment also increases the agitation underpinning social warming: nothing contributes to low-level anger quite like an unequal application of the rules. The idea that, as Clegg suggested in a speech in September 2019, Facebook offers a tennis court where it rolls the surface, paints the lines and checks the net, but 'we don't pick up a racket and start playing,' doesn't stand up to cursory examination.[13] The umpire is biased.

Until the end of 2020, none of the platforms had any significant reason to fix the problem; there was no existential threat. If Isis or al-Qaida or white supremacists organised and recruited on one or the other or all of the platforms, wasn't that just like the old days, where they could gather in forums and spread ideas? (That, you'll recall, was Monika Bickert's assertion.) The difference, however, is that the platforms are so much bigger, and fuelled by algorithms, which together means those network effects, and the geometric growth of problems, are inevitable.

Before considering how we can fix the problem, it's worth considering how to classify it. Is this like global warming, which is simultaneously gigantic and made up of billions of tiny personal choices? Tackling that demands changes in our daily behaviour that seem too far-reaching to contemplate, and improvement seems far off on a decades-long trajectory. That would imply that fixing the problem of social warming requires us all mindfully to use the platforms less, to be aware of their calls to outrage and to resist them.

Or might this be a problem akin to the ozone hole? That was also caused by the individual choices of huge numbers of people – yet the damage was reversed by regulations imposed in the 1980s by governments on the small number of companies manufacturing pollutants called CFCs, used in hairsprays and refrigerants.

Certainly, you could comfortably fit the chief executives of all the social networking companies into the average boardroom. In 2020, there were fifteen social networks with more than 300 million users, serving a total user audience of 3.8 billion, 99 percent of whom accessed those networks via their smartphone.[14] Ranked by size, the biggest platforms in the West are Facebook (taking in Facebook itself, WhatsApp, Facebook Messenger and Instagram, each with more than a billion users), YouTube, TikTok, Reddit, Snapchat, Twitter and Pinterest. There are some that are effectively exclusive to China: WeChat, Douyin (the Chinese form of TikTok), QQ, Qzone, Sina Weibo and Kuaishou.

There's substantial overlap in user bases: six networks each had more than a billion users. Size doesn't necessarily correlate with public influence or influence on social warming – Reddit has 100 million more users than Twitter, which is only just larger than Pinterest, but only one of those three garners constant media attention.

The social networks are akin to the producers of CFCs: without them, the problems we are seeing would be minimal.

And just as CFCs went from miracle compound to chemical villain in a few short years, despite rearguard PR campaigns by their manufacturers, so a growing number of people who were positive about social networks in the past have soured on them. There are the former employees such as YouTube's Chaslot or Facebook's Zhang. Or there are those such as Roger McNamee, a long-time venture capitalist who was a personal early investor in Facebook itself. In 2012, he told Bloomberg that 'I think Mark Zuckerberg is "The One". Like Bill Gates and Steve Jobs, he has set a tone that everyone else has lined up behind.'[15] Four years later, having watched aghast the manipulation and exploitation of the site in the US election by malicious groups, he became one of the company's – and Zuckerberg's – most prominent critics:

'Internet platforms are causing great harm and [we should] accept that executives like Mark Zuckerberg are not sincere in their promises to do better,' he asserted in June 2020.[16] The mission of 'connecting everyone' had become an overriding dogma, ignoring the harms of harassment, purposeful disinformation and conspiracy theories, he explained.

Others went further: Facebook is 'destroying the very fabric of democracy, destroying the very fabric of human relationships and peddling in an addictive drug called anger', said Joanna Hoffman, one of the members of Apple's original Macintosh project and a former adviser to Steve Jobs, at a conference in June 2020.[17] 'You know, it's just like tobacco, it's no different than the opioids. We know anger is addictive, we know we can attract people to our platform and get engagement if we get them pissed off enough. So what, we should capitalise on that each and every time?' Hoffman accused young technology leaders of being 'remarkably ignorant on what they are sowing in the world', and wondered about their motives: 'The question is, how flawed, how ignorant, and how devious?'

Or take Yaël Eisenstat, the former CIA officer who quit Facebook in frustration six months after being hired to head its 'election integrity operations'. Speaking in 2020, she suggested Facebook would never address the fundamental flaws that allowed the manipulation that threatened democracies: it was too scared of angering potential regulators, and the separate desire to make everything scale meant the detailed attention that different countries demanded would not be provided. Shira Ovide, a journalist at the *New York Times*, commented in June 2020: 'Every time I think something positive about Facebook, I also hold in my mind the profound damage the company has done – and that too should be an indelible part of Zuckerberg's legacy.'[18] Having a site where people could put videos of police brutality, such as the killing of George Floyd, was important. 'But we shouldn't accept a genocide in Myanmar or the targeting of a journalist in the Philippines in exchange for it.'

Even Mark Zuckerberg managed to half-admit there was a problem, announcing in October 2020 that content that 'denies or distorts' the reality of the Holocaust would be prohibited, saying that his thinking had 'evolved' since he had said in an interview in 2018 that 'it's hard to impugn intent' regarding those who indulged in denial. The reason he had changed his mind, he said, was that the company's policies on hate speech had evolved too, and 'I've seen data showing an increase in anti-Semitic violence'. How much, people wondered, had Facebook's laxity contributed to that rise in violence? Neither Facebook nor Zuckerberg would say. It was, however, one small step against social warming from its foremost protagonist.

So: there is a problem, and lots of people, even including those who run the companies, recognise at least some of what's going wrong.

Logically, then, what we need is regulation. What aspect of the platforms do we regulate, though? We can't uninvent them. And despite the fulminations of the right wing in the US, forcing the sites to be liable for everything that appears on them – in effect repealing Section 230 of the Communications Decency Act – is impracticable.

Before trying to frame new laws, what about self-regulation? There have been intermittent signs that the platforms are taking their effects on their users seriously. In summer 2018, Instagram introduced a measure that indicated when you'd seen all the new posts added in the past forty-eight hours, or since you'd last used the app. 'You're All Caught Up', a little onscreen message announced. Kevin Systrom, the co-founder and at the time CEO, had tweeted in May that year that 'any time [spent on Instagram] should be positive and intentional'[19] and added that 'understanding how time online impacts people is important, and it's the responsibility of all companies to be honest about this. We want to be part of the solution. I take that responsibility seriously.'[20] In 2019, the number of Likes on a post was made less visible. Positive moves, if you're trying to reduce that itch to compare yourself against others.

That all looked great – until in mid-2020 Instagram undid the first part of its fix. If you scrolled back past the point of being Caught Up, you'd see posts from 'suggested' accounts to follow, algorithmically picked of course. Systrom had left and Instagram was under new, attention-hungry management.

Facebook, meanwhile, belatedly made a number of changes towards the end of 2020. Groups being used by white supremacists (such as those Julia Carrie Wong had pointed out) were formally banned from the site. The QAnon ban was followed by Zuckerberg's reversal on Holocaust denial, which had previously been allowed (with Zuckerberg, in that 2018 interview, saying he thought people should be allowed to be 'wrong' about topics – revealing a failure to understand the purpose of Holocaust denialists). Then came a ban on adverts that discouraged people from getting vaccinated (although adverts advocating for or against legislation around vaccines would be allowed, as would the many Groups pushing anti-vaccine rhetoric and misinformation). Twitter, meanwhile, tried to stop people reflexively retweeting articles without reading them by putting up a dialog pointing out that 'Headlines don't tell the full story' and encouraging people to read them first. Retweeting also became fractionally more difficult: instead, people were encouraged to quote tweet, adding commentary to the original.

This was encouraging. But it didn't get to the root of the problem. In 2014, Twitter's moderators essentially stood aside as Gamergate ripped apart some people's lives. In 2015 and 2016, the biggest problem on YouTube, Twitter and Facebook seemed to be Islamic extremist content, and the networks acted against them with determination. Yet at the same time they overlooked or ignored the rapid rise of alt-right content, Russian disinformation, the bubbling anger in Myanmar, anti-vaccine misinformation and the use of fake accounts to influence opinion. The problem is that though QAnon and Holocaust denial might be the topics that needed proactive moderation by the end of 2020 (even though in both cases action came too late), that doesn't tell us what the problem topics will be in one or two years' time. All we can be certain about is that unless there are important changes, there will be problems just as big, just as pervasive.

Naughton thinks that in seeking to control the discourse of their users, the platforms may be trying to attempt something that can't be done. He quotes Ashby's Law of Requisite Variety, derived by a British cybernetician in 1954. 'The implications of it are that for a system to be viable – any system – it has to be able to cope with the complexity of its environment. And if it can't do that, it isn't viable. Mass production is one example, when you say, like Henry Ford, that you can have your car in any colour you want, as long as it's black. That's an example of variety reduction. You reduce the complexity of the environment to a level that you can handle. In other cases, you ramp up your internal capacity to handle complexity by making the organisation more flexible or more reactive. There are various ways to do it.

'Now, if you look at, for example, what's happening with the tech companies. What they're saying is, "Well, yes it's very difficult but we'll find a way of dealing with the complexity thrown at us by 2.8 billion people." How will they do it? Well, first off they're spending a lot of money on having moderators observe user engagement. But even if Facebook had half a million moderators, it's not doable.' Because humans are so diverse, the complexity of problems that the real world is throwing at Facebook and other social networks requires them to be able to respond with as much nuance – or 'variety' – as the problems that people generate. But with billions of people in hundreds of countries using different languages with varying cultures and histories, that is a colossal task; even saying that hate speech will be removed is too challenging.

'So because of Ashby's Law, though they don't talk in these terms, they're basically saying, "We're going to amplify our variety internally by using AI,"' Naughton says. 'That's their bet at the moment, My hunch is that that won't work because any kind of conceivable AI is not going to be smart enough to do it. So it's doomed.'

Doomed in the sense that Facebook will collapse through its own contradictions?

'Doomed in the sense that society gets destroyed by it,' Naughton says.

In spring 2020, the American Economic Liberties Project, a think tank, published a working paper that asked: what if Facebook and Google didn't exist, and someone proposed giving effective control over global discussions and knowledge to two corporations?[21] One would be a platform for fake news stories, enthusiastically amplified, and for 'individually tailored' propaganda that could influence elections and even (in the pandemic) affect public health. The other would intimately surveil adults and children, use its designs to keep users hooked, enable subtle forms of discrimination through targeted advertising, and separately drive the collapse of journalism, both nationally and locally.

Put like that, they don't sound very desirable. But that's what we have, the authors argued, suggesting that Facebook and Google should be broken up, and new regulations introduced to stop them becoming concentrated again.

As you might imagine, the companies are a little wary of being scaled down. 'Breaking up these companies, whether it's Facebook or Google or Amazon, is not actually going to solve the issues,' Zuckerberg told staff in an all-hands meeting in October 2019.[22]

Doing so, he argued, would reduce their resources and prevent them coordinating to prevent election interference and hate speech.

This was a little disingenuous. The criticism of Facebook and YouTube particularly is that their gigantic scale enables malicious actors to reach more people at once; in the days when there were dozens of social networks, any disinformation scheme would have had to do far more work to reach the same number of people as can be reached by a single Facebook advertising campaign now. Nothing would have precluded the networks from coordinating to spot such exploitation, though.

All of this points to the first possible solution: restrict the size of the networks.

We already have an indicator that this works. When Twitter permanently bans individuals such as Katie Hopkins or Alex Jones or David Duke or Bill Mitchell (right-wing agitprops seem to figure heavily on the list collected at Wikipedia), they storm off to smaller networks and are effectively never heard from again by the wider world.[23] Donald Trump's abrupt removal from Twitter and Facebook wasn't quite as effective, but even he was struck silent for days by his banishment, and eschewed other outlets because they lacked the same reach. We can remain uninformed and essentially unworried by what is happening on the many other social networks besides Facebook, Instagram or Twitter – the smaller ones such as Gab, Parler, CounterSocial, Mastodon, Pinterest, Telegram, Taringa! (in Latin America), Tumblr – indicating one important way in which social warming differs from global warming or the ozone hole. If you have half as many cars, you have half the global warming effect. But cutting the size of a network by half dramatically reduces the size of potential interactions. Facebook is clearly too big, unable to handle the problems that its own scale and laissez-faire attitude have created. If it were forced to split into multiple separate companies, each limited to, say, 250 million users at most, just about enough to cover North America (its most valuable territory), then it should be able to control the content on the site, while using its better knowledge of those users to sell advertising.[24] The world could be full of little unconnected Facebooks, mostly at a national level, forced to be separate. Facebook, Instagram, WhatsApp and Twitter wouldn't like it. But everyone else would probably benefit. Such a measure would echo the legislation already in place in almost every country limiting the concentration of media ownership, which implicitly recognises that too much concentration of power over what people are told is dangerous to governments and citizens. The social networks are not publishers; but they can relentlessly amplify the worst of society, just as monopolistic publishers can.

Smaller networks would also be better positioned to check against dangerous and misleading adverts. Anne Borden King, who with her ad-hoc group of colleagues has fought against misinformation on vaccines and autism, told me that 'Ads you would never hear on the radio or see on the television are flying across social media without action by regulators or watchdogs.' The reliance on AI to mediate adverts is a huge problem, she says, as it obscures lines of responsibility and allows huge numbers of harmful adverts to slip through. 'If they really want to do it, to stop the steady flow of Covid misinformation and other pseudoscience, they could end it tomorrow by stopping the use of AI that doesn't work and using people to review the ads instead.' Humans given a strict set of guidelines would be far better at the job, she says – which, given what we know about the limitations of today's AI, is entirely true.

What about the expense of checking the ads? 'Not doing what's right because it's less profitable is not a reasonable excuse,' King retorts. 'Losing money by doing the right thing is a decision many of us make every day of our lives. No one who is hauled before the courts for tax evasion or Ponzi schemes can make the excuse:

"Oh, but if I stopped doing it, I wouldn't be able to make as much money."'

We also know that the companies do take notice when governments act. In Germany, social media companies were asked in 2015 to remove hate speech from their platform within twenty-four hours, as part of a self-regulation process. That didn't happen. So in June 2017, the German government passed the *Netzwerkdurchsetzungsgesetz*, or Network Enforcement Act, which pointed out that German penal law, rather than Facebook's Community Standards, decided what would be allowed on the network in Germany. Failure to remove hate speech within twenty-four hours and illegal content within seven days would incur a fine of up to €50 million.[25]

Within a year, one in six of Facebook's moderators were working in Germany.

Other changes could also have a significant effect. Roger McNamee suggests that Section 230 of the CDA should be tweaked to allow people to sue platforms for the algorithmic amplification of harmful content. That would leave sites free to moderate as they wished, but make them liable for the way in which they presented and chose content. 'Harmful amplification is the result of business choices that can be changed,' he observed.

Such a sweeping change would, if it had been in force previously, have made Facebook open to lawsuits over events in the Philippines election, put YouTube on notice for the radicalisation of a number of fanatical Islamic and right-wing killers, and made Facebook take more notice of the activity inside closed Groups where people were plotting attacks against the police and others. Ordinary forums that present content in chronological order wouldn't be affected, and would still have the protection of Section 230. But those reliant on algorithmic amplification would have to consider their choices carefully. It's worth remembering that all of the platforms have had a period without algorithmic amplification; Twitter and Instagram only adopted it in 2016.

Even if McNamee's idea isn't adopted, the pressure towards regulation has intensified as the power of Google and especially Facebook has become clearer. The US senator Elizabeth Warren is insistent that 'Today's big tech companies have too much power – too much power over our economy, our society and our democracy.'[26] That isn't a partisan position: a poll carried out in October 2020, in nine American 'battleground' states ahead of the election, found 74 percent of voters would support a government-led break-up of big tech companies 'to reduce their power and increase competition and innovation'.[27]

Craig Silverman, at Buzzfeed News, thinks that Facebook and the others might welcome some sort of clear legal framework for their content, which they mostly have to make up on their own at present. 'I think they realise that they don't really want all of the power they have, the control and responsibility they have,' he told me. 'They would love to be able to hand off the really tough decisions to someone else, and be able to just run their business. The idea that regulation might actually alleviate some of the pressure and responsibility suddenly became a little more appealing – like "Man, it would be great if we could just kick some stuff to other people and make them responsible for it, wouldn't that be good, and then we could just sell ads, you know, that'd be great."'

There's another change that would positively affect the broader experience of the networks: use fewer of the onscreen signals and systems that we know contribute to the problems. The ideal would make it a little more difficult to do things that lead to warming, while not taking away the benefits of being able to contact friends and to create and discover content.

You'd want a design where the temptation to insult and annoy and hassle whomever you like can't easily be satisfied. We already know that outrage has a strong tendency to go viral, and that we notice content with emotional or moral words more than 'neutral' phrasings. A more beneficial social network would make it harder to pass on others' content, or to comment on it in the way Twitter's quote tweet does. It wouldn't gamify our experiences by encouraging us to compare ourselves against others; instead, rather like Ben Grosser's Demetricator browser add-on, it would hide details such as how popular a post is, or how many followers a person has, so that we would make the decision on whether to follow them purely on the quality of their content, not their apparent fame.

To Grosser's quiet satisfaction, social networks began in 2019 to take some of the numbers off their displays. That April, Adam Mosseri at Instagram told developers that follower counts would be 'less prominent' on users' profiles, and that there wouldn't be a specific count for the number of Likes under the posts of people you followed. Twitter and YouTube made similar noises, and though Twitter didn't follow through, YouTube did stop updating subscriber numbers in real time, and rounded the figure that was displayed, thus reducing fights between professional YouTubers over who was 'biggest'.

Sometimes, of course, you do need outrage in order to battle injustices. The explosion of anger in the US in mid-2020 over the death of George Floyd was only made possible by the ubiquity of smartphones, and of data networks, and the fact that Facebook permitted users to upload and distribute videos (even if they showed people being mortally harmed). The outrage became the point. A few years ago, before almost anyone could afford a smartphone and a data plan, before social networks were everywhere, the police brutality was the same, the deaths as numerous. But to show them to the wider world you'd have to attract the attention of a TV station or newspaper. The police would have reported a death from 'existing health issues', and there might barely have been a mention in the local newspaper; what, an editor would have said, is the *story*?

That's not to say it would be impossible. The 1992 Los Angeles riots were triggered by the acquittal of four police officers accused of the beating of Rodney King – events that were caught on film and televised, as were many brutal incidents in the riots themselves. Sometimes the outrage that we encounter online is necessary to improve the world. The astonished fury that people felt at seeing Facebook Live streams or Twitter videos of unarmed black men and women being killed by American police was an eye-opener. Social warming can even, *in extremis*, be virtuous.

'Social media is good at taking people down, bringing things down, deconstructing them,' Alan Davis, the international observer who watched Myanmar's already serious ethnic tensions become more and more inflamed by Facebook use there, told me. 'It's not good at building them up. It's not good at solutions.'

Yet we know what the problem is, and we know the distortions of the social networks that make them happen: their scale, the algorithms, the ease of sharing content.

There's nothing axiomatic about any of those. Limiting the size of networks works to reduce their impact. Social systems existed without using algorithms to amplify content; forums around the web still function today. And the ease with which content can be shared is often inversely proportional to how much time we spend reflecting on whether we should share it, and what benefit it might bring. 'Engagement' is a misleading metric, in thrall to advertising; when used by the military, it means conflict. And sometimes social networks look like warfare by means of text, pictures and videos.

Legislators have the power to make changes. Forcing social networks to limit their size and algorithmic amplification would not be popular – especially with the fantastically rich people who own them – but we can glimpse the benefits by thinking back a few years to a time before they dominated, when we were not constantly fretting about their effects on society.

The solutions, then, are up to us. We cannot make social networks disappear, and many of us wouldn't want to. We need instead to be aware of how we can extract the most value from them. We are social animals, and all the advances that have brought us to the intensely interdependent civilisations that we live in now have been achieved by collective action. The times in human history when our species has been most imperilled have been when we are divided or ranged against each other. Social networks keep promising to unite us, yet their design works in the opposite direction. They have become a tool that does not do what we want.

In the past, our toolmaking ancestors knew what to do with something that didn't work as required: redesign and reshape it to match your desired outcome. We can't change our essential natures. But we can change our tools. The extent of social warming demonstrates that the time has come to reshape and redesign these broken tools we have come to rely on.

ACKNOWLEDGEMENTS

The first inklings of this idea came through the work I did at Cambridge University's Centre for Technology and Democracy in 2016, looking at the polarising effects of time spent online. I'm grateful to Professor John Naughton, who gave me the opportunity there to begin the research which eventually became this book. Thanks too to my agent, Doug Young, who listened to the much more boring book topic that I pitched to him, looked at the ceiling for a bit and said, 'Do you have any other ideas?' Which led me to offer this one as Plan B, which became Plan A.

Thank you to all the team at Oneworld, particularly my editor Cecilia Stein for her patient reading and excellent suggestions about changes, as well as Rida Vaquas, Juliana Pars, Margot Weale, Paul Nash and Jacqui Lewis, who checked names and corrected all the grammar which . . . that was wrong.

There are of course so many people who have helped and supported in so many ways. A special mention must go to Shian Powell for all the cakes, and Becky McGrath for making me work off the effects of the cakes.

A final word of gratitude must go to Jojo, for her enormous reserves of kindness, generosity and patience. I couldn't have done any of this without her.

NOTES

All website links were accurate at the time of writing (January 2021).

PROLOGUE: THE SHAPE OF THE PROBLEM

1 european-rhetoric.com/analyses/ikeynote-analysis-iphone/transcript-2007
2 techcrunch.com/2008/04/29/end-of-speculation-the-real-twitter-usage-numbers
3 theverge.com/2017/2/16/14642164/facebook-mark-zuckerberg-letter-mission-statement
4 wearesocial.com/blog/2018/01/global-digital-report-2018
5 thesun.co.uk/sport/football/12101685/wilfried-zaha-racist-abuse-crystal-palace
6 sanfrancisco.cbslocal.com/2020/06/17/steven-carrillo-robert-alvin-justis-facebook-boogaloo-crackdown
7 wearesocial.com/blog/2018/01/global-digital-report-2018; statista.com/statistics/617136/digital-population-worldwide
8 todayinsci.com/D/Diesel_Rudolf/DieselRudolf-DieselEngine.htm

EARLY DAYS: THE PROMISE AND THE POWER

1 thoughtco.com/henry-ford-biography-1991814
2 wired.com/1997/05/ff-well
3 angelfire.com/bc3/dissident
4 h2o.law.harvard.edu/cases/4540
5 law.cornell.edu/uscode/text/47/230
6 sifry.com/alerts/2006/08/state-of-the-blogosphere-august-2006
7 web.archive.org/web/20010202020100/rebeccamead.com/2000_11_13_art_blog.htm
8 fawny.org/decon-blog.html

9 medium.com/@worstonlinedater/tinder-experiments-ii-guys-unless-you-are-really-hot-you-are-probably-better-off-not-wasting-your-2ddf370a6e9a
10 web.archive.org/web/20100309222302/foresight.org/Updates/Update02/Update02.3.html
11 archive.fortune.com/magazines/fortune/fortune_archive/2003/10/13/350905/index.htm
12 eonline.com/news/506119/mark-zuckerberg-talks-facebook-s-10th-anniversary
13 nytimes.com/2015/11/11/arts/international/rene-girard-french-theorist-of-the-social-sciences-dies-at-91.html
14 twitter.com/jack/status/20
15 reuters.com/article/us-myspace-idUSTRE7364G420110407
16 ft.com/content/fd9ffd9c-dee5-11de-adff-00144feab49a
17 pcworld.com/article/162719/social_networks.html
18 news.bbc.co.uk/1/hi/programmes/click_online/5391258.stm
19 theguardian.com/media/pda/2009/nov/24/future-of-social-networks-twitter-linkedin-mobile-application-next
20 web.archive.org/web/20090227043048/blog.facebook.com/blog.php?post=56566967130
21 web.archive.org/web/20111207043554/http:/nytimes.com/2008/08/03/magazine/03trolls-t.html?hp=&pagewanted=all
22 researchgate.net/publication/8451443_The_Online_Disinhibition_Effect
23 web.archive.org/web/20131015201115/http:/nbcnews.com/id/26837911/ns/health-behavior/t/anonymity-opens-split-personality-zone
24 techcrunch.com/2009/03/07/eric-schmidt-tells-charlie-rose-google-is-unlikely-to-buy-twitter-and-wants-to-turn-phones-into-tvs
25 ted.com/talks/amber_case_we_are_all_cyborgs_now/transcript
26 technosociology.org/?p=102
27 twitter.com/FawazRashed/status/48882406010257408
28 globalvoices.org/2007/08/30/arabeyes-who-is-using-the-tunisian-presidential-airplane
29 portland-communications.com/publications/the-arab-spring-and-the-future-of-communications
30 theguardian.com/world/2011/feb/25/twitter-facebook-uprisings-arab-libya

31 United Nations Conference on Trade and Development, *ICT Policy Review*: Egypt, Switzerland, United Nations Publication, 2011, p. 2. United Nations Conference on Trade and Development, 2011, p. xiii.
32 Merlyna Lim, 'Clicks, Cabs, and Coffee Houses: Social Media and Oppositional Movements in Egypt, 2004–2011,' *Journal of Communication*, vol. 62, no. 2, 2012, pp. 231–248 and cspo.org/legacy/library/1207150932F24192826YK_lib_LimJoC2012Egypt.pdf
33 jstor.org/stable/pdf/10.7249/j.ctt4cgd90.10.pdf
34 portland-communications.com/publications/the-arab-spring-and-the-future-of-communications
35 bbc.co.uk/blogs/newsnight/paulmason/2011/02/twenty_reasons_why_its_kicking.html
36 reuters.com/article/us-technology-risk/insight-social-media-a-political-tool-for-good-or-evil-idUSTRE78R3CM20110929
37 paleycenter.org/assets/international-council/IC-2011-LA/Beyond-Disruption-Lo-Res-2011.pdf
38 money.cnn.com/2011/11/08/technology/zuckerberg_charlie_rose/index.htm
39 technologyreview.com/2011/08/23/117825/streetbook
40 ofcom.org.uk/__data/assets/pdf_file/0013/20218/cmr_uk_2012.pdf
41 Ofcom 2012, p. 221 – 50% takeup; p. 35 – one-third use daily
42 ofcom.org.uk/about-ofcom/latest/features-and-news/uk-now-a-smart-phone-society
43 pewresearch.org/global/2012/12/12/social-networking-popular-across-globe

AMPLIFICATION AND ALGORITHMS: THE WATCHER BENEATH YOUR SCREEN

1 wired.com/2012/06/google-x-neural-network
2 facebook.com/careers/life/the-impact-of-machine-learning-at-facebook-community-integrity-and-innovation
3 blog.twitter.com/en_us/a/2016/increasing-our-investment-in-machine-learning.html
4 computerworld.com/article/3086179/heres-why-twitter-bought-machine-learning-startup-magic-pony.html
5 amazon.com/Making-Fly-Genetics-Animal-Design/dp/0632030488

6 michaeleisen.org/blog/?p=358
7 patents.google.com/patent/US9110953B2/en?q=(newsfeed+popularity+facebook)&assignee=Facebook%2c+Inc.&before=filing:20100101&after=filing:20090101
8 web.archive.org/web/20060911084122/blog.facebook.com/blog.php?post=2207967130
9 danah.org/papers/FacebookAndPrivacy.html
10 marketingland.com/edgerank-is-dead-facebooks-news-feed-algorithm-now-has-close-to-100k-weight-factors-55908
11 twitter.com/kevinroose/status/1306678570576523264
12 pnas.org/content/111/24/8788.full
13 medium.com/message/engineering-the-public-289c91390225
14 theguardian.com/technology/2014/jun/29/facebook-users-emotions-news-feeds
15 washingtonpost.com/news/morning-mix/wp/2014/06/30/facebook-responds-to-criticism-of-study-that-manipulated-users-news-feeds
16 ideas.ted.com/need-to-know-about-facebooks-emotional-contagion-study
17 arxiv.org/pdf/1803.03453.pdf
18 money.cnn.com/2017/08/17/technology/culture/facebook-hate-groups/index.html
19 buzzfeednews.com/article/craigsilverman/how-facebook-groups-are-being-exploited-to-spread
20 about.fb.com/news/2018/04/keeping-terrorists-off-facebook
21 whistleblowers.org/wp-content/uploads/2019/05/Facebook-SEC-Petition-2019.pdf
22 whistleblowers.org/wp-content/uploads/2019/05/Facebook-SEC-Petition-2019.pdf (p.11 footnote)
23 washingtonpost.com/news/the-switch/wp/2018/04/11/transcript-of-zuckerbergs-appearance-before-house-committee
24 telegraph.co.uk/news/2018/05/05/facebook-accused-introducing-extremists-one-another-suggested
25 fsi-live.s3.us-west-1.amazonaws.com/s3fs-public/stamos_written_testimony_-_house_homeland_security_committee_-_ai_and_counterterrorism.pdf
26 apnews.com/f97c24dab4f34bd0b48b36f2988952a4
27 apnews.com/3479209d927946f7a284a71d66e431c7

28 reuters.com/article/us-facebook-boogaloo/facebook-moves-to-limit-spread-of-boogaloo-groups-after-charges-idUSKBN23C011
29 menendez.senate.gov/news-and-events/press/menendez-colleagues-blast-facebooks-inaction-to-prevent-white-supremacist-groups-from-using-its-platform-as-a-recruitment-and-organizational-tool
30 allthingsd.com/20131104/in-hatching-twitter-a-billion-dollar-company-that-almost-wasnt
31 web.archive.org/web/20160315223405/blog.instagram.com/post/141107034797/160315-news
32 engadget.com/2017/08/08/instagram-algorithm
33 slate.com/articles/technology/cover_story/2017/03/twitter_s_timeline_algorithm_and_its_effect_on_us_explained.html
34 twitter.com/maplecocaine/status/1080665226410889217
35 techcrunch.com/2011/06/19/youtube-counts-video-ads-regular-views
36 youtube-creators.googleblog.com/2012/08/youtube-now-why-we-focus-on-watch-time.html
37 web.archive.org/web/20110202122438/lri.fr/~teytaud/mogo.html
38 storage.googleapis.com/pub-tools-public-publication-data/pdf/45530.pdf
39 youtube.com/yt/about/press
40 fool.com/investing/2018/02/06/people-still-spend-an-absurd-amount-of-time-on-fac.aspx
41 nytimes.com/2018/03/10/opinion/sunday/youtube-politics-radical.html
42 medium.com/@MediaManipulation/unite-the-right-how-youtubes-recommendation-algorithm-connects-the-u-s-far-right-9f1387ccfabd
43 youtube.googleblog.com/2017/07/bringing-new-redirect-method-features.html
44 medium.com/@guillaumechaslot/how-youtubes-a-i-boosts-alternative-facts-3cc276f47cf7
45 journals.sagepub.com/doi/pdf/10.1177/0392192116669288
46 youtube.googleblog.com/2019/01/continuing-our-work-to-improve.html
47 threadreaderapp.com/thread/1150486778090086400.html
48 farid.berkeley.edu/downloads/publications/arxiv20.pdf and nytimes.com/interactive/2020/03/02/technology/youtube-conspiracy-theory.html
49 arxiv.org/pdf/1902.10730v1.pdf
50 twitter.com/DeepMind/status/1101514121563041792
51 nytimes.com/2019/06/03/world/americas/youtube-pedophiles.html

52 theverge.com/2017/11/15/16656706/youtube-videos-children-comments
53 youtube.googleblog.com/2017/11/5-ways-were-toughening-our-approach-to.html

OUTRAGE AND SCISSOR STATEMENTS: OUR TRIBAL MINDSET

1 ncbi.nlm.nih.gov/pmc/articles/PMC2504725
2 scn.ucla.edu/pdf/fairness_nzherald.pdf
3 psychology.yale.edu/news/molly-crockett-join-psychology-department-faculty-2017
4 doi:10.1038/s41562 and static1.squarespace.com/static/538ca3ade4b090f9ef331978/t/5a53c0d49140b7212c35b20e/1515438295247/Crockett_2017_NHB_Outrage.pdf
5 researchgate.net/publication/265606809_Morality_in_everyday_life
6 nytimes.com/2020/02/21/us/white-supremacist-guilty-of-killing-2-who-came-to-aid-of-black-teens.html
7 archive.indianexpress.com/news/what-akhilesh-yadav-wont-tell-inaction-administrative-failure-in-muzaffarnagar/1167448/0
8 archive.indianexpress.com/news/what-akhilesh-yadav-wont-tell-inaction-administrative-failure-in-muzaffarnagar/1167448/0
9 archive.indianexpress.com/news/muzaffarnagar-rioters-used-whatsapp-to-fan-flames-find-police/1168072/0
10 firstpost.com/india/pune-muslim-techie-killed-by-rightwing-mob-over-morphed-fb-posts-1555709.html
11 web.archive.org/web/20140702194511/jana.com/blog/facebook-can-juice-whatsapp-billions-from-emerging-markets
12 wsj.com/articles/the-internet-is-filling-up-because-indians-are-sending-millions-of-good-morning-texts-1516640068
13 whatsapp.com/research/awards
14 blog.whatsapp.com/more-changes-to-forwarding
15 downloads.bbc.co.uk/mediacentre/duty-identity-credibility.pdf
16 indianexpress.com/article/cities/delhi/northeast-delhi-riots-whatsapp-group-6488320
17 newslaundry.com/2020/07/04/delhi-riots-inside-the-kattar-hindu-whatsapp-group-that-planned-executed-murders

18 cjr.org/tow_center/whatsapp-doesnt-have-to-break-encryption-to-beat-fake-news.php
19 economictimes.indiatimes.com/news/politics-and-nation/lynch-mobs-are-an-indian-problem-not-whatsapps/articleshow/65034017.cms
20 web.archive.org/web/20141218033302/slatestarcodex.com/2014/12/17/the-toxoplasma-of-rage
21 nytimes.com/2020/05/30/opinion/sunday/trump-twitter-jack-dorsey.html
22 bengrosser.com/projects/facebook-demetricator
23 rhizome.org/editorial/2012/nov/15/dont-give-me-numbers-interview-ben-grosser-about-f
24 computationalculture.net/what-do-metrics-want
25 medium.com/message/my-name-is-danah-and-im-a-stats-addict-93f7636320bb
26 theguardian.com/uk-news/2017/jul/13/viscount-jailed-for-offering-money-for-killing-of-gina-miller
27 blog.twitter.com/official/en_us/a/2009/retweet-limited-rollout.html
28 buzzfeednews.com/article/alexkantrowitz/how-the-retweet-ruined-the-internet
29 deadspin.com/the-future-of-the-culture-wars-is-here-and-its-gamerga-1646145844
30 researchgate.net/publication/314092336_Measuring_GamerGate_A_Tale_of_Hate_Sexism_and_Bullying
31 medium.com/message/72-hours-of-gamergate-e00513f7cf5d
32 presenttensejournal.org/wp-content/uploads/2018/06/Trice_Potts.pdf
33 theatlantic.com/magazine/archive/2018/04/the-case-against-retweets/554078
34 quora.com/How-many-users-did-Twitter-have-after-its-first-year
35 medium.com/@Luca/how-to-turn-off-retweets-for-everyone-99dd835c10f8
36 twitter.com/luca/status/974190006729412608
37 theringer.com/tech/2018/5/2/17311616/twitter-retweet-quote-endorsement-function-trolls
38 psyarxiv.com/3n9u8
39 wsj.com/articles/facebook-knows-it-encourages-division-top-executives-nixed-solutions-11590507499

40 vulture.com/2017/08/the-toxic-drama-of-ya-twitter.html
41 newyorker.com/books/under-review/in-ya-where-is-the-line-between-criticism-and-cancel-culture
42 twitter.com/smorganspells
43 twitter.com/briandavidearp/status/1090350858989195265
44 medium.com/@monteiro/twitters-great-depression-4dc394ed10f4
45 web.stanford.edu/~gentzkow/research/facebook.pdf

WORST-CASE SCENARIO: HOW FACEBOOK SENT MYANMAR HAYWIRE

1 data.worldbank.org/indicator/IT.CEL.SETS.P2
2 economist.com/asia/2014/03/28/too-much-information
3 web.archive.org/web/20090429182629/minorityrights.org/3546/briefing-papers/minorities-in-burma.html
4 economist.com/banyan/2014/09/04/the-leftovers
5 linkedin.com/pulse/rise-anti-muslim-hate-speech-shortly-before-outbreaks-nickey-diamond
6 ericsson.com/assets/local/about-ericsson/sustainability-and-corporate-responsibility/documents/download/communication-for-all/myanmar-report-2012-13nov.pdf
7 data.worldbank.org/country/myanmar
8 iflr.com/pdfs/newsletters/Asia%20Telecoms%20-%20Myanmar-an%20untapped%20telco%20market%20(14%20Mar%202012).pdf
9 gsma.com/publicpolicy/wp-content/uploads/2012/11/gsma-deloitte-impact-mobile-telephony-economic-growth.pdf
10 themimu.info/sites/themimu.info/files/documents/Presentation_Why_Stop_Zawgyi_Use_Unicode_Phandeeyar_Aug2016.pdf
11 frontiermyanmar.net/en/features/battle-of-the-fonts
12 irrawaddy.com/news/4-killed-24-injured-as-buddhists-and-muslims-clash-in-central-burma.html
13 academia.edu/9587031/New_Technologies_Established_Practices_Developing_Narratives_of_Muslim_Threat_in_Myanmar
14 foxnews.com/tech/google-chairman-urges-myanmar-to-embrace-free-speech-and-private-sector-telecom-development
15 aljazeera.com/programmes/101east/2013/09/201393134037347273.html

16 Melissa Crouch (ed) *Islam and the State in Myanmar: Muslim–Buddhist Relations and the Politics of Belonging* global.oup.com/academic/product/islam-and-the-state-in-myanmar-9780199461202
17 iri.org/sites/default/files/wysiwyg/public_opinion_survey_burma_june-july_2019_english.pdf
18 cbsnews.com/news/rohingya-refugee-crisis-myanmar-weaponizing-social-media-main/; adb.org/sites/default/files/publication/176518/ewp-462.pdf
19 pri.org/stories/2014-08-08/newly-liberated-myanmar-hatred-spreads-facebook
20 wired.com/story/how-facebooks-rise-fueled-chaos-and-confusion-in-myanmar
21 refworld.org/docid/559bd57b28.html
22 news.un.org/en/story/2014/07/473952-myanmar-un-rights-expert-warns-against-backtracking-free-expression-association
23 wired.com/story/how-facebooks-rise-fueled-chaos-and-confusion-in-myanmar
24 adb.org/sites/default/files/linked-documents/49470-001-so.pdf
25 ifes.org/sites/default/files/Myanmar-2015-2016-National-Election-Survey.pdf
26 reuters.com/investigates/special-report/myanmar-facebook-hate
27 statista.com/statistics/489191/average-daily-cellular-usage-by-country
28 nytimes.com/2015/07/20/world/asia/those-who-would-remake-myanmar-find-that-words-fail-them.html
29 web.archive.org/web/20160607094422/ericsson.com/res/docs/2016/mobility-report/emr-raso-june-2016.pdf
30 Suspicious Minds: user perceptions of privacy on Facebook in Myanmar lirneasia.net/wp-content/uploads/2018/07/ITS2018_Hurulle_Suspicous-minds_paper.docx
31 mmtimes.com/national-news/21787-how-social-media-became-myanmar-s-hate-speech-megaphone.html
32 mmtimes.com/business/technology/20816-facebook-racks-up-10m-myanmar-users.html
33 iri.org/sites/default/files/8.25.2017_burma_public_poll.pdf
34 pbs.org/wgbh/frontline/film/facebook-dilemma/transcript
35 rayms.github.io/2018-05-09-twitter-s-myanmar-hate-machine
36 theguardian.com/world/2018/apr/03/revealed-facebook-hate-speech-exploded-in-myanmar-during-rohingya-crisis

37 buzzfeednews.com/article/sheerafrenkel/fake-news-spreads-trump-around-the-world
38 uk.reuters.com/article/us-myanmar-rohingya-facebook/u-n-investigators-cite-facebook-role-in-myanmar-crisis-idUKKCN1GO2PN
39 bbc.co.uk/news/technology-43385677
40 bbc.co.uk/news/technology-43385677
41 reuters.com/investigates/special-report/myanmar-facebook-hate
42 buzzfeednews.com/article/meghara/facebook-myanmar-rohingya-genocide
43 about.fb.com/news/2018/11/myanmar-hria
44 about.fb.com/wp-content/uploads/2018/11/bsr-facebook-myanmar-hria_final.pdf
45 data.worldbank.org/indicator/NY.GDP.MKTP.CD?end=2018&locations=MM&most_recent_value_desc=false&start=2000&type=points&view=chart&year=2003
46 ericsson.com/en/blog/2016/10/making-solar-power-economically-viable-for-our-children

DIVIDED VOTE: HOW SOCIAL MEDIA POLARISES POLITICS

1 statista.com/statistics/515599/connected-services-whatsapp-usage-penetration-uk
2 standard.co.uk/comment/comment/leaving-politics-was-gutwrenching-enough-without-being-booted-out-of-the-tory-mps-whatsapp-group-a4314551.html
3 theguardian.com/politics/blog/2009/jul/29/cameron-swearing-interview
4 twitter.com/David_Cameron/status/254625004321386496
5 thetimes.co.uk/article/brexit-backing-mps-plot-their-attacks-on-whatsapp-sw5gp7680
6 skwawkbox.org/2020/04/12/leaked-document-accuses-senior-right-wing-labour-staff-working-against-corbyn-and-reveals-their-dismay-at-electoral-surge
7 papers.ssrn.com/paper.taf?abstract_id=199668
8 buzzfeed.com/alexspence/these-leaked-whatsapp-chats-reveal-just-how-brexiteer
9 markpack.org.uk/2450/another-twitter-first

10 theguardian.com/business/2012/nov/21/wonga-apologises-stella-creasy-abusive-twitter-messages
11 pewresearch.org/fact-tank/2017/08/21/highly-ideological-members-of-congress-have-more-facebook-followers-than-moderates-do
12 people-press.org/2017/02/23/partisan-conflict-and-congressional-outreach
13 twitter.com/patel4witham/status/1200792218749022213
14 twitter.com/TonyKent_Writes/status/1200819605243138050
15 ribbonfarm.com/2016/09/15/crowds-and-technology
16 fowler.ucsd.edu/massive_turnout.pdf
17 politico.com/news/2019/10/30/facebook-twitter-political-ads-062297
18 bloombergquint.com/opinion/the-2-8-million-non-voters-who-delivered-brexit
19 bbc.co.uk/news/uk-politics-44966969
20 lordashcroftpolls.com/2019/02/how-the-uk-voted-on-brexit-and-why-a-refresher
21 buzzfeed.com/laurasilver/the-tories-are-buying-google-ads-to-squeeze-the-lib-dems-in
22 wired.com/2016/11/facebook-won-trump-election-not-just-fake-news
23 theguardian.com/us-news/2016/dec/09/trump-and-clintons-final-campaign-spending-revealed
24 npr.org/2016/12/06/504520364/how-trump-waged-an-under-the-radar-ground-game
25 washingtonpost.com/politics/how-brad-parscale-once-a-nobody-in-san-antonio-shaped-trumps-combative-politics-and-rose-to-his-inner-circle/2018/11/09/b4257d58-dbb7-11e8-b3f0-62607289efee_story.htmlf
26 buzzfeednews.com/article/ryanhatesthis/mueller-report-internet-research-agency-detailed-2016
27 digitalmarketingcommunity.com/researches/paid-media-benchmark-report-q4-2019-adstage
28 c-span.org/video/?465293-1/facebook-ceo-testimony-house-financial-services-committee at 4'18' and cnet.com/news/facebook-rejected-biden-request-to-pull-false-trump-ad-about-ukraine
29 bbc.co.uk/news/technology-42102570
30 washingtonpost.com/technology/2020/06/28/facebook-zuckerberg-trump-hate

31 adl.org/news/article/sacha-baron-cohens-keynote-address-at-adls-2019-never-is-now-summit-on-anti-semitism
32 politico.com/news/2020/09/26/facebook-conservatives-2020-421146
33 pnas.org/content/pnas/117/1/243.full.pdf
34 onezero.medium.com/russian-trolls-arent-actually-persuading-americans-on-twitter-study-finds-d8fd6bcacaba
35 stanford.edu/~dbroock/published%20paper%20PDFs/kalla_broockman_minimal_persuasive_effects_of_campaign_contact_in_general_elections_evidence_from_49_field_experiments.pdf
36 gq.com/story/russian-trolls-targetting-black-voters
37 warner.senate.gov/public/_cache/files/5/7/57d1657c-7e9c-46ce-be61-8a4ed1499e2c/D4801DCE15F2026882E1D6219F00DFB7.10.6.20-facebook-honest-ads-letter.docx.pdf
38 warner.senate.gov/public/_cache/files/0/2/0283bfb8-93f0-42e4-9a7d-dc4864e51547/A027982BADA652B832485E16640641A9.10.6.20-google-honest-ads-act-letter.docx.pdf
39 warner.senate.gov/public/_cache/files/3/a/3afc73bd-d03f-43be-801d-85417c6c55e6/0589911AC5097909F38E0FA5B772FEB2.10.6.20-twitter-honest-ads-act-letter.docx.pdf

FACT TO FAKE: HOW THE MEDIA ECOSYSTEM COLLAPSED

1 twitter.com/jkrums/status/1121915133
2 cnbc.com/2014/01/15/the-five-year-anniversary-of-twitters-defining-moment.html
3 edition.cnn.com/2009/TECH/01/22/social.networking.news
4 arstechnica.com/tech-policy/2009/06/twitter-from-statedept-delay-upgrade-to-aid-iran-protests
5 web.archive.org/web/20090619022828/news.sky.com/skynews/Home/World-News/Iran-Election-Coverage-On-Twitter-Social-Media-Becomes-Vital-Tool-For-Iranian-Citizens/Article/200906315307209
6 web.archive.org/web/20000131092232/shirky.com/writings/information_price.html
7 web.archive.org/web/20140331235717/shirky.com/writings/paying_attention.html
8 journalism.org/fact-sheet/newspapers

9 web.archive.org/web/20051211131216/fortune.com/fortune/technology/articles/0,15114,1135386,00.html

10 statista.com/statistics/266206/googles-annual-global-revenue

11 daringfireball.net/thetalkshow/2020/03/26/ep-280

12 digiday.com/media/facbeook-twitter-ferguson

13 cjr.org/tow_center_reports/the_traffic_factories_metrics_at_chartbeat_gawker_media_and_the_new_york_times.php

14 riaa.com/u-s-sales-database

15 buzzfeednews.com/article/mathonan/why-facebook-and-mark-zuckerberg-went-all-in-on-live-video

16 youtube.com/watch?v=iROUSLUUgzE 2'05

17 techcrunch.com/2015/11/04/facebook-video-views

18 facebook.com/business/news/facebook-video-metrics-update

19 mercurynews.com/2019/10/07/facebook-agrees-to-pay-40-million-over-inflated-video-viewing-times-but-denies-doing-anything-wrong

20 slate.com/technology/2018/10/facebook-online-video-pivot-metrics-false.html

21 buzzsumo.com/blog/facebook-video-engagement-learned-analyzing-100-million-videos

22 theonion.com/christ-article-a-video-1819595653

23 theatlantic.com/technology/archive/2018/10/facebook-driven-video-push-may-have-cost-483-journalists-their-jobs/573403

24 usnewsdeserts.com/reports/expanding-news-desert/loss-of-local-news/loss-newspapers-readers

25 facebook.com/business/news/news-feed-fyi-bringing-people-closer-together

26 nytimes.com/2016/10/31/business/media/publishers-rethink-outbrain-taboola-ads.html

27 forbes.com/sites/kalevleetaru/2019/03/04/visualizing-seven-years-of-twitters-evolution-2012-2018

28 techonomy.com/conf/te16/videos-conversations-with-2/in-conversation-with-mark-zuckerberg

29 theguardian.com/technology/2016/nov/10/facebook-fake-news-election-conspiracy-theories

30 buzzfeednews.com/article/craigsilverman/top-fake-news-of-2016

31 ipsos.com/en-us/news-polls/ipsosbuzzfeed-poll-fake-news

32 niemanlab.org/2016/05/pew-report-44-percent-of-u-s-adults-get-news-on-facebook

33 news.gallup.com/poll/1597/confidence-institutions.aspx
34 pewresearch.org/internet/fact-sheet/social-media
35 facebook.com/zuck/posts/a-lot-of-you-have-asked-what-were-doing-about-misinformation-so-i-wanted-to-give/10103269806149061
36 facebook.com/formedia/blog/working-to-stop-misinformation-and-false-news
37 theguardian.com/technology/2017/mar/22/facebook-fact-checking-tool-fake-news
38 buzzfeednews.com/article/brookebinkowski/fact-checking-facebook-doomed
39 about.fb.com/news/2017/09/information-operations-update
40 facebook.com/zuck/posts/10104067130714241
41 techcrunch.com/2017/09/27/mark-zuckerberg-responds-to-anti-trump-claim
42 buzzfeednews.com/article/craigsilverman/these-are-50-of-the-biggest-fake-news-hits-on-facebook-in
43 buzzfeednews.com/article/craigsilverman/facebook-just-shared-the-first-data-about-how-effective-its
44 researchgate.net/publication/321887941_The_Implied_Truth_Effect_Attaching_Warnings_to_a_Subset_of_Fake_News_Headlines_Increases_Perceived_Accuracy_of_Headlines_Without_Warnings
45 dailycaller.com/2018/03/02/facebook-snopes-fact-checks-demonitize-christian-satire
46 archive.is/snopes.com/fact-check/cnn-washing-machine
47 poynter.org/fact-checking/2018/snopes-is-feuding-with-one-of-the-internets-most-notorious-hoaxers
48 politifact.com/article/2017/may/31/If-youre-fooled-by-fake-news-this-man-probably-wro
49 politifact.com/factchecks/2017/may/15/blog-posting/liberal-troll-behind-tucker-carlson-death-hoax-spr
50 poynter.org/fact-checking/2017/a-satirical-fake-news-site-apologized-for-making-a-story-too-real
51 buzzfeednews.com/article/craigsilverman/facebook-fake-news-hits-2018
52 psyarxiv.com/9qdza
53 bbc.co.uk/news/technology-47098021
54 poynter.org/fact-checking/2019/snopes-pulls-out-of-its-fact-checking-partnership-with-facebook

55 snopes.com/disclosures
56 web.archive.org/web/20170709135912/changeadvertising.org/the-click-bait-report
57 nytimes.com/2016/10/31/business/media/publishers-rethink-outbrain-taboola-ads.html
58 wired.co.uk/article/fake-news-outbrain-taboola-hillary-clinton
59 digiday.com/media/shift-publishers-can-no-longer-count-content-recommendation-guarantee-checks
60 gmfus.org/blog/2020/10/12/new-study-digital-new-deal-finds-engagement-deceptive-outlets-higher-facebook-today
61 washingtonpost.com/blogs/erik-wemple/wp/2017/09/08/politico-editor-we-discard-dozens-of-potential-hires-over-toxic-twitter-feeds
62 edition.cnn.com/2020/06/07/us/pittsburgh-newspaper-black-journalist-looting-tweet/index.html
63 int.nyt.com/data/documenthelper/7010-recommendations-for-social-med/a5c91e59333f4fa0c8bf/optimized/full.pdf#page=1
64 washingtonpost.com/lifestyle/style/washington-post-suspends-reporter-who-tweeted-about-kobe-bryant-rape-allegations-following-his-death/2020/01/27/babe9c04-413b-11ea-b5fc-eefa848cde99_story.html
65 yorkshireeveningpost.co.uk/news/people/it-was-chaos-shocking-photo-shows-leeds-four-year-old-suspected-pneumonia-forced-sleep-floor-lgi-due-lack-beds-1334909
66 theguardian.com/media/2019/dec/10/woman-says-account-hacked-to-post-fake-story-about-hospital-boy
67 twitter.com/tnewtondunn/status/1204077603356852225
68 firstdraftnews.org/latest/how-two-disinformation-campaigns-swung-into-action-days-before-the-uk-goes-to-the-polls
69 twitter.com/davidyelland/status/1204299182691037185
70 twitter.com/bbclaurak/status/1204110491242643457 and twitter.com/tnewtondunn/status/1204097378170626048
71 newhistories.group.shef.ac.uk/wordpress/wordpress/the-1992-election-the-nhs-and-the-%e2%80%98war-of-jennifer%e2%80%99s-ear%e2%80%99
72 twitter.com/JayMitchinson/status/1204344653174181888/photo/1
73 thetimes.co.uk/edition/news/emily-maitlis-bbc-hires-ex-news-chief-to-ensure-stars-stay-impartial-on-social-media-kbd0qsxnt

74 thetimes.co.uk/article/bbc-journalists-addicted-to-toxic-twitter-says-boss-flxgzr6k3#:~:text=Some%20BBC%20journalists%20have%20become,use%20the%20%E2%80%9Ctoxic%E2%80%9D%20platform

DEMOCRACY AT RISK: WHY SOCIAL MEDIA UNDERMINES ELECTIONS

1 npr.org/2011/02/16/133775340/twitters-biz-stone-on-starting-a-revolution?t=1593517727343
2 newsbusters.org/blogs/nb/kyle-gillis/2010/07/22/facebook-ceo-defends-company-abc-interview-advocates-democracy
3 fredturner.stanford.edu/wp-content/uploads/Turner-Trump-on-Twitter-in-PJB-ZZP.pdf
4 washingtonpost.com/technology/2020/06/28/facebook-zuckerberg-trump-hate
5 brown.edu/Research/Shapiro/pdfs/cross-polar.pdf
6 digitalsocietyproject.org/wp-content/uploads/2019/05/DSP_WP_01-Introducing-the-Digital-Society-Project.pdf
7 theconversation.com/why-the-oromo-protests-mark-a-change-in-ethiopias-political-landscape-63779
8 advox.globalvoices.org/2018/01/20/leaked-documents-show-that-ethiopias-ruling-elites-are-hiring-social-media-trolls-and-watching-porn
9 hrw.org/news/2020/03/09/ethiopia-communications-shutdown-takes-heavy-toll
10 abcnews.go.com/International/wireStory/ethiopias-pm-warns-longer-internet-cutoff-amid-unrest-64712610
11 advox.globalvoices.org/2019/08/07/in-ethiopia-disinformation-spreads-through-facebook-live-as-political-tensions-rise
12 nobelprize.org/prizes/peace/2019/abiy/109716-lecture-english
13 technology.inquirer.net/8013/more-filipinos-now-using-internet-for-news-information-study
14 rappler.com/life-and-style/technology/41959-globe-free-facebook-access
15 prnewswire.com/news-releases/facebook-ceo-mark-zuckerberg-philippines-a-successful-test-bed-for-internetorg-initiative-with-globe-telecom-partnership-247184981.html
16 web.archive.org/web/20160313143504/rappler.com/brandrap/profile-internet-users-ph

17 news.abs-cbn.com/business/01/25/17/filipinos-lead-the-world-in-social-media-use-survey
18 advox.globalvoices.org/2017/07/27/can-facebook-connect-the-next-billion
19 advox.globalvoices.org/2017/07/28/philippines-on-facebooks-free-version-fake-news-is-even-harder-to-spot
20 facebook.com/papalvisitph/photos/a.321295728024881.1073741828.321279424693178/566912616796523/?type=3
21 newmandala.org/how-duterte-won-the-election-on-facebook
22 rappler.com/newsbreak/investigative/fake-accounts-manufactured-reality-social-media
23 rappler.com/nation/148007-propaganda-war-weaponizing-internet
24 edition.cnn.com/2020/06/30/opinions/maria-ressa-facebook-intl-hnk/index.html
25 vox.com/world/2018/10/29/18025066/bolsonaro-brazil-elections-voters-q-a
26 public.tableau.com/profile/zeroratingcts#!/vizhome/zeroratinginfo/Painel1
27 nytimes.com/2019/08/11/world/americas/youtube-brazil.html
28 nytimes.com/2019/08/15/the-weekly/how-youtube-misinformation-resolved-a-whatsapp-mystery-in-brazil.html
29 arxiv.org/abs/1804.00397
30 buzzfeednews.com/article/ryanhatesthis/brazils-congressional-youtubers
31 newsweek.com/jailed-politician-leads-polls-brazils-presidential-election-1081643
32 bbc.co.uk/news/world-latin-america-45380237
33 aaai.org/ojs/index.php/ICWSM/article/view/3271
34 nytimes.com/2018/10/17/opinion/brazil-election-fake-news-whatsapp.html
35 bnnbloomberg.ca/facebook-whatsapp-step-up-efforts-in-brazil-s-fake-news-battle-1.1157003
36 theguardian.com/world/2019/oct/30/whatsapp-fake-news-brazil-election-favoured-jair-bolsonaro-analysis-suggests
37 washingtonpost.com/world/the_americas/brazil-bolsonaro-fake-news-coronavirus/2020/06/03/60194428-a4de-11ea-898e-b21b9a83f792_story.html
38 folha.uol.com.br/opiniao/2018/10/como-o-whatsapp-combate-a-desinformacao-no-brasil.shtml (tr Google Translate)

39 theconversation.com/whatsapp-skewed-brazilian-election-proving-social-medias-danger-to-democracy-106476
40 atlanticcouncil.org/wp-content/uploads/2020/06/operation-carthage-002.pdf
41 medium.com/dfrlab/facebook-disabled-assets-linked-to-egypt-and-uae-based-firms-a232d9effc32
42 timeslive.co.za/news/south-africa/2017-09-04-how-the-gupta-campaign-weaponised-social-media
43 dc.sourceafrica.net/documents/118115-Manufacturing-Divides.html
44 timeslive.co.za/news/south-africa/2017-09-04-analysis--so-did-the-gupta-familys-fake-news-campaign-work and dc.sourceafrica.net/documents/118115-Manufacturing-Divides.html
45 washingtonmonthly.com/2006/09/25/did-he-or-didnt-he
46 youtube.com/watch?v=dJ1wcpsOtS4
47 edition.cnn.com/2020/06/30/opinions/maria-ressa-facebook-intl-hnk/index.html
48 edition.cnn.com/2019/07/09/media/maria-ressa-media-press-freedom-conference-intl/index.html
49 theguardian.com/technology/2020/jul/26/yael-eisenstat-facebook-is-ripe-for-manipulation-and-viral-misinformation
50 buzzfeednews.com/article/craigsilverman/facebook-ignore-political-manipulation-whistleblower-memo
51 cigionline.org/articles/maria-ressa-facebook-broke-democracy-many-countries-around-world-including-mine

PANDEMIC: INOCULATED AGAINST TRUTH

1 un.org/en/un-coronavirus-communications-team/un-tackling-‘infodemic’-misinformation-and-cybercrime-covid-19
2 uk.finance.yahoo.com/news/mark-zuckerberg-touts-broad-power-174405399.html
3 blog.twitter.com/en_us/topics/company/2020/covid-19.html
4 web.archive.org/web/20200131070919/about.fb.com/news/2020/01/coronavirus
5 hughhewitt.com/mark-zuckerberg-on-facebook-shops-a-challenge-to-amazon-presidents-xi-and-trump-and-ccp-disinformation-fbs-oversight-board-amd-joe-bidens-dislike-of-him
6 thelancet.com/pdfs/journals/lancet/PIIS0140-6736(20)30461-X.pdf

7 pubmed.ncbi.nlm.nih.gov/27776823
8 nytimes.com/2019/08/15/the-weekly/how-youtube-misinformation-resolved -a-whatsapp-mystery-in-brazil.html
9 wired.com/story/the-rise-and-spread-of-a-5g-coronavirus-conspiracy -theory
10 twitter.com/NWOforum/status/1223870765893857281
11 usafacts.org/visualizations/coronavirus-covid-19-spread-map; twitter.com/ fbnewsroom/status/1239703497479614466
12 msn.com/en-us/news/world/china-fuels-coronavirus-conspiracy-theory -blaming-us-army/ar-BB117UEx
13 web.archive.org/web/20200317200104/blog.twitter.com/en_us/topics/ company/2020/An-update-on-our-continuity-strategy-during-COVID -19.html
14 about.fb.com/wp-content/uploads/2020/03/March-18-2020-Press-Call-Transcript.pdf
15 huffingtonpost.co.uk/entry/david-clarke-twitter-coronavirus_n_ 5e7008d3c5b60fb69ddc921d?ri18n=true
16 theverge.com/2020/3/16/21181617/twitter-coronavirus-misinformation -tweets-removed-accounts-covid19
17 twitter.com/elonmusk/status/1240753430001356801
18 twitter.com/elonmusk/status/1240754657263144960
19 twitter.com/elonmusk/status/1240758710646878208
20 nytimes.com/2020/03/17/health/coronavirus-childen.html
21 theverge.com/2020/3/20/21187760/twitter-elon-musk-tweet-coronavirus -misinformation
22 techcrunch.com/2020/03/25/twitter-federalist-coronavirus
23 twitter.com/TwitterSafety/status/1240418440982040579
24 theverge.com/2020/3/30/21199845/twitter-tweets-brazil-venezuela-presidents-covid-19-coronavirus-jair-bolsonaro-maduro
25 researchgate.net/publication/321887941
26 pewresearch.org/fact-tank/2020/04/08/nearly-three-in-ten-americans-believe-covid-19-was-made-in-a-lab
27 theguardian.com/technology/2020/apr/06/at-least-20-uk-phone-masts-vandalised-over-false-5g-coronavirus-claims
28 reutersinstitute.politics.ox.ac.uk/types-sources-and-claims-covid-19 -misinformation
29 blog.whatsapp.com/Keeping-WhatsApp-Personal-and-Private

30 techcrunch.com/2020/04/27/whatsapps-new-limit-cuts-virality-of-highly-forwarded-messages-by-70
31 twitter.com/FoldableHuman/status/1244361971572428800
32 facebook.com/zuck/posts/10111806366438811
33 themarkup.org/coronavirus/2020/04/23/want-to-find-a-misinformed-public-facebooks-already-done-it
34 s21.q4cdn.com/399680738/files/doc_financials/2019/q4/Q4-2019-Earnings-Presentation-_final.pdf
35 doi.org/10.1177/0392192116669288
36 kclpure.kcl.ac.uk/portal/en/publications/the-relationship-between-conspiracy-beliefs-and-compliance-with-public-health-guidance-with-regard-to-covid19(734ca397-6a4d-4208-bc1a-f3da12f04628).html
37 searchenginejournal.com/google-facebook-ban-ads-for-face-masks-as-coronavirus-spreads/354402
38 bbc.co.uk/news/technology-52517797
39 buzzfeednews.com/article/janelytvynenko/coronavirus-plandemic-viral-harmful-fauci-mikovits
40 theguardian.com/world/2017/feb/14/bill-gates-philanthropy-warren-buffett-vaccines-infant-mortality
41 buzzfeed.com/cameronwilson/coronvirus-antivaxxers-facebook-instagram-boost
42 facebook.com/communitystandards/recentupdates/regulated_goods
43 books.google.co.uk/books/about/Healing_the_Symptoms_Known_As_Autism.html?id=OwibnQEACAAJ
44 wsls.com/health/2019/05/21/meet-the-moms-who-are-battling-droves-of-parents-who-believe-bleach-will-cure-their-childs-autism
45 counterhate.co.uk/anti-vaxx-industry
46 telegraph.co.uk/technology/2020/10/14/facebook-ban-anti-vaccination-adverts
47 nature.com/articles/s41586-020-2281-1.pdf
48 cdc.gov/nchs/fastats/immunize.htm
49 humanetech.com/about-us
50 twitter.com/alistaircoleman/status/1336388520827613188
51 youtube.com/watch?v=E5yyInwI7tw

REGULATION: CUTTING THE PROBLEM DOWN TO SIZE

1 washingtonpost.com/archive/politics/1988/04/10/cfcs-rise-and-fall-of-chemical-miracle/9dc7f67b-8ba9-4e11-b247-a36337d5a87b

2 businessinsider.com/mark-zuckerberg-interview-with-axel-springer-ceo-mathias-doepfner-2016-2

3 zephoria.com/top-15-valuable-facebook-statistics

4 zephoria.com/twitter-statistics-top-ten

5 about.fb.com/news/2020/07/facebook-does-not-benefit-from-hate

6 theguardian.com/commentisfree/2020/jul/01/facebook-hate-speech-policy-advertising

7 news.yahoo.com/fbi-documents-conspiracy-theories-terrorism-160000507.html

8 platformer.news/p/why-did-facebook-ban-qanon-now

9 zephoria.com/twitter-statistics-top-ten

10 theverge.com/2018/10/24/18018486/twitter-presence-indicators-activity-status-ice-breakers

11 forbes.com/sites/carlieporterfield/2020/06/03/twitter-suspends-account-copying-trumps-tweets-for-glorifying-violence

12 nytimes.com/2020/06/03/technology/facebook-trump-employees-letter.html

13 about.fb.com/news/2019/09/elections-and-political-speech

14 smartinsights.com/social-media-marketing/social-media-strategy/new-global-social-media-research

15 bloomberg.com/news/articles/2012-05-17/how-mark-zuckerberg-hacked-the-valley

16 time.com/5855733/social-media-platforms-claim-moderation-will-reduce-harassment-disinformation-and-conspiracies-it-wont

17 cnbc.com/2020/06/12/joanna-hoffman-facebook-is-peddling-an-addictive-drug-called-anger.html

18 nytimes.com/2020/06/16/technology/facebook-philippines.html

19 twitter.com/kevin/status/996543082899226624

20 twitter.com/kevin/status/996543333097865216

21 economicliberties.us/our-work/addressing-facebook-and-googles-harms-through-a-regulated-competition-approach

22 theverge.com/2019/10/1/20892354/mark-zuckerberg-full-transcript-leaked-facebook-meetings

23 en.wikipedia.org/wiki/Twitter_suspensions
24 statista.com/statistics/247614/number-of-monthly-active-facebook-users-worldwide
25 bbc.co.uk/news/technology-42510868
26 medium.com/@teamwarren/heres-how-we-can-break-up-big-tech-9ad9e0da324c
27 publicpolicypolling.com/wp-content/uploads/2020/10/Senate-Battleground-Oct-2020-Results.pdf

Charles Arthur is a journalist, author and speaker, writing on science and technology for over thirty years. He was technology editor of the *Guardian* from 2005–2014, and afterwards carried out research into social division at Cambridge University. He is the author of two specialist books, *Digital Wars* and *Cyber Wars*.